Thermal Processing of Hydrocarbons

THERMAL PROCESSING OF HYDROCARBONS

PETROLEUM TO PETROCHEMICALS

DWIJEN K. BANERJEE

PennWell®

Disclaimer

The recommendations, advice, descriptions, and the methods in this book are presented solely for educational purposes. The author and publisher assume no liability whatsoever for any loss or damage that results from the use of any of the material in this book. Use of the material in this book is solely at the risk of the user.

Endeavor Business Media, LLC
1421 South Sheridan Road
Tulsa, Oklahoma 74112-6600 USA

800.752.9764
+1.866.777.1814
sales@pennwell.com
www.pennwellbooks.com

Publisher: Matthew Dresher

Cover image: © Getty Images - imantsu

Library of Congress Cataloging-in-Publication Data

Names: Banerjee, Dwijen K., author.
Title: Thermal processing of hydrocarbons : petroleum to petrochemicals / Dwijen Banerjee.
Description: First edition. | Tulsa, Oklahoma, USA : PennWell, 2019. | Includes bibliographical references and index.
Identifiers: LCCN 2019050342 (print) | LCCN 2019050343 (ebook) | ISBN 9781593702656 (hardcover) | ISBN 9781593702779 (ebook)
Subjects: LCSH: Pyrolysis. | Petroleum—Refining. | Petroleum products—Heating.
Classification: LCC TP156.P9 B36 2019 (print) | LCC TP156.P9 (ebook) | DDC 665.5—dc23
LC record available at https://lccn.loc.gov/2019050342
LC ebook record available at https://lccn.loc.gov/2019050343

Printed in the United States of America

1 2 3 4 5 23 22 21 20 19

This book is dedicated to my late parents:
They chose to sacrifice and invest their last penny in me and my siblings only to teach us the importance of education. Since my early childhood, I pledged that I will make something out of this. I went to one of the top schools in India, studied in Europe, and received their highest educational degrees. I started an amazing professional journey in a turbulent energy industry in the 1980s, navigated through the 1990s and 2000s, worked in Canada and the US, travelled to many countries throughout the world, and learned something new every day.

Contents

Chapter 4

Chapter 5

Chapter 6

Chapter 7

Chapter 8

Preface

It was the summer solstice—a sizzling day when the temperature hovered between double and triple digits according to my car dashboard. Listening to Chopin while driving with my wife, Anca, a classical pianist, I was heading from Oklahoma to the Atlantic coast, almost 1,000 miles away. We were traveling to Georgia to visit our son Christopher, a resident physician training to become a neurosurgeon. The summer heat made me curious to ask my son about how the body and brain respond to and adapt to the highest heat for the longest period of time during the solstice. I wondered if such questions were addressed during his medical school training in Miami, where the climate was commonly the etiology of "heat strokes."

My pleasant daydreaming shifted as along the highway, I saw endless fragments of carbon strips from the bursting of huge tires from the heavy transportation trucks. Because of my profession, I questioned how the heat might play a role in their demise, as I knew that tires are basically made of carbon bonded together with sulfur as

...C-C-S-C-S-S-C-C.....

and that these bonds break down under the pressure of heat, high speed, and the frictional forces of tires on the asphalt—another carbon product. A similar phenomenon takes place when hydrocarbons are subjected to heat, pressure, and a catalyst: the bigger-sized molecules break down into smaller hydrocarbons through the rupture of their C-C bonds. Being the longest day of the year, in scientific terms, that translates into the environment getting the maximum amount of heat for the longest period of time, i.e., the two necessary factors that are required for any thermal reactions to take place. Thinking about my son's training and now about a topic I am much more familiar with, I recalled my own training and expertise in this field and reminded myself of the importance of quality education. Having already written a book, *Oil Sands, Heavy Oil, and Bitumen*, I now decided to write a second and continue to pass forward my own knowledge and four decades of experience in the petroleum industry.

My connection to the thermal processing of hydrocarbons goes back to mid-1970s, when I started working on a second doctoral program at the Institute of Petroleum and Gas in Romania (my first PhD was from the Indian Institute of Technology, Delhi), where my research focused on the pyrolysis of aromatic hydrocarbons. After I completed my studies, I continued to work on "fast-pyrolysis" with then world-renowned professor Keith Laidler at the University of Ottawa, Canada. Dr. Laidler collaborated in the study of kinetics and thermodynamics with other world-renowned scientists, including Henry Eyring (nominated for a Nobel Prize), Sir Cyril Hinshelwood (awarded a knighthood by the Queen of England), and Canadian Nobel Laureate John Polanyi. I am honored to be part of an admirable lineage of researchers in this area (please see chart below).

This book organizes and summarizes research on the thermal processing of hydrocarbons over the past half-century. Because it has been done by scientists all over the world, I cannot individually thank everyone; however, I would like to acknowledge the variety of scientific contributors on this subject, whose research findings I am sharing with you in the following chapters. The many ideas expressed by the various schools of thoughts on such a complex topic make it almost impossible for everyone to agree on everything or to draw definitive conclusions. This is science; there are diverse opinions, and where debate continues, I am not expecting it to stop here in this book.

The application of thermal processing of hydrocarbons goes back to the early 19th century, when lighter gaseous hydrocarbons were produced from liquid hydrocarbons for use in European street lights. During the First World War, the application of thermal cracking of crude oil was used for a darker purpose: in the then Soviet Union, toluene was used as feedstock for producing the explosive Trinitro-Toluene (TNT).

This book will help R & D engineers in the energy business, graduate students, chemists, and petrochemical scientists understand the basic principles of the thermal processing of hydrocarbons. The first two chapters explain, in simple language, the physics and chemistry of thermal processes. This especially narrows the gap between the chemical engineers and chemists. Later chapters deal with various commercial applications of thermal processing of petroleum hydrocarbons. The different sections of this book can be read separately.

Recently, because of a worldwide increase in demand for plastics, there has been a surge in the proposals for mega projects like building integrated refineries and petrochemicals. This book can help operational and processing engineers and scientists trying to rush to adopt these changes in the refineries.

Another industry accelerating at high speed, especially in North America, is the production of shale gas/shale oil. The production of these hydrocarbons, also known as tight gas/tight oil, are due to the development of a new technology "hydraulic fracturing," commonly called "fracking." This book therefore includes a chapter on tight oil.

Nearly seventy-five years ago, Professor Wilbur L. Nelson wrote his classic book *Petroleum Refining*. It was like a bible to me while I was a graduate student, and in his preface, he says,

"An engineer must spend more time applying information than deriving or finding it ... the current literature is so voluminous that it may be useless to a busy engineer unless it is presented in an organized form." This is even more true today. Given the recent explosion in the tight oil and petrochemical industries, I want to follow in Dr. Nelson's footsteps. Instead of providing voluminous references and conflicting opinions, I have kept this information as simple and organized as possible, so it can be useful to today's busy engineers and scientists. I am sure this book is timely for the events of today.

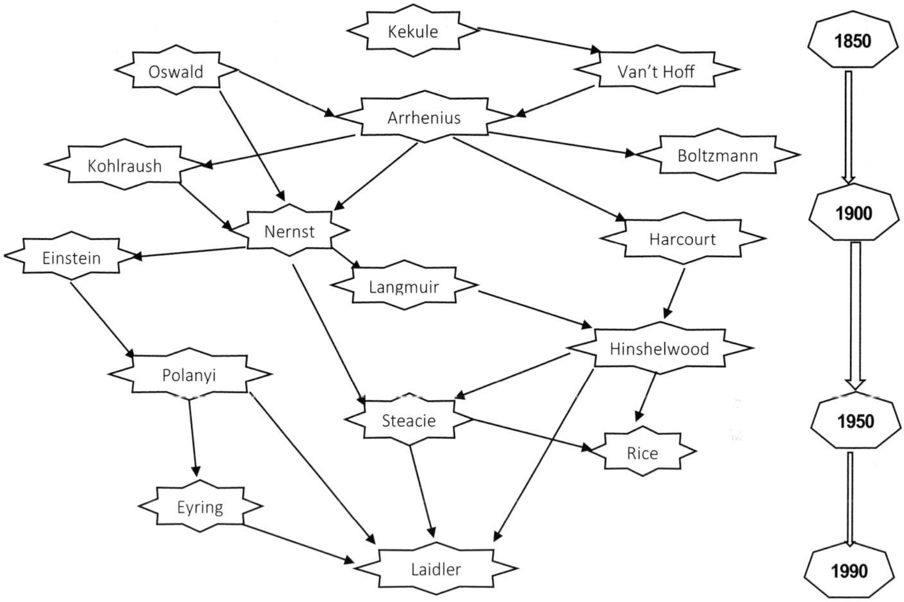

The History of Kinetics and Thermodynamics

(The arrows indicate collaborated with or worked under)

This chart was presented by the late Prof. Laidler in a meeting in Ottawa, Canada in 1982, while I was working with him. The years shown on the right hand side of the chart are especially added here for the readers to understand the time line involved.

1

Composition of hydrocarbons in crude oil

1.1 Introduction

In the petroleum industry, the hydrocarbon series could be divided into three major sections as follows:

i. Natural gas
ii. Liquid hydrocarbons or crude oil
iii. Extra heavy oil and bitumen (almost solid at room temperature)

Although coal is a type of hydrocarbon that falls into a fourth category as part of the energy industry, it is a rock solid carbonaceous material, the dirtiest of all the fossil fuels, which is in the process of being replaced by the relatively cleaner fossil fuel, natural gas, for the sake of environmental benefit.

1.2 Natural gas

The hydrocarbon series begins with the lowest hydrocarbons C_1 to C_4 that fall in the natural gas ranges. The natural gas that people are usually familiar with mostly consists of methane or CH_4. But when natural gas is recovered from underground reservoirs, it contains paraffinic gaseous hydrocarbons in the range from C_1 to C_4, along with some liquid C_5+ hydrocarbons, with a decrease in concentrations with the increase in molecular weight; methane being in the highest concentrations of more than 75 vol% and pentane less than 1 vol%. Natural gas also contains inorganic gases such as hydrogen sulfide and carbon dioxide. When exposed, hydrogen sulfide is extremely toxic and can kill operators working in the field. The composition of natural gas varies from reservoir to reservoir.

If the natural gas contains enough condensable material such as C_3 to C_5 hydrocarbons at a concentration above 0.05 liter per meter cube, it is called "wet gas" and it is quite valuable for the industry. The hydrocarbons are recovered and condensed under pressure, then sold as liquefied natural gas (LNG) in the market.

Due to environmental concerns and also the increase in natural gas production from a new technology called fracking (see later in Chapter 5), the global LNG market is rapidly increasing. The success of LNG production is thus dependent on selecting the most cost-effective technology that can liquefy and separate C_3–C_5 hydrocarbons from C_1–C_2 dry gas (see Chapter 5 on tight oil for details). Heavier hydrocarbons are also removed from the dry natural gas stream by simple refrigeration before transportation, to avoid plugging of the pipeline and to maintain BTU value of the dry gas stream.

While there is an ongoing debate in the industry about environmental concerns and the future of fossil fuel, it is true that natural gas is the solution for a lower carbon future to meet both ends, i.e., energy and the environment. The market demand shows that there is strong growth of natural gas all over the world. According to the Oil & Gas Journal (July 2018), the world consumed 3% more gas last year at almost twice the historical rate of growth. Further, natural gas consumption in the US increased by 7% in 2018, and by about 10% in the power generation industry alone.

The demand for natural gas also increased in another sector—the petrochemical industry (see Chapter 6 for detail), as there are several products depending on it as feedstock: isoprene, butadiene, resins, synthetic rubber, polyethylene, polypropylene, etc.

1.3 Crude oil

The next in the series is C_5 to C_{50} hydrocarbons, usually liquid at normal conditions and falling within the crude oil group. The hydrocarbons become more complex as the carbon number in the series progressively increases from C_1 to C_{50} and beyond. Composition of crude oil varies drastically from well to well, reservoir to reservoir, place to place, country to country. Most commonly, it is characterized in the oil industry by the density and the boiling point range. However, the oil industry prefers to use its own density scale called Degree API, which is actually an expanded scale and the inverse of specific gravity. Degree API is defined by the equation:

$$\text{Degree API Gravity} = \frac{141.5}{\textit{Specific Gravity}\,@15.6C} - 131.5$$

According to the American Petroleum Institute classification, crude oil is divided into various classes based on its °API as follows:

Table 1.3.1. Classification of crude oil

Crude type	°API range
Light crude	>31.1
Medium crude	22.3 – 31.1
Heavy crude	10.0 – 22.3
Extra Heavy / Bitumen	<10.0

Each type of crude oil is further classified based on sulfur content: if the concentration is below 1.0 wt%, it is called "sweet" crude and if above, it is called "sour" crude. Crude oil is valued in the market based on its degree API and sulfur content.

Major crudes worldwide in the light and medium ranges are located mainly in the European, Asian, and African regions. These crude oils are recovered by conventional methods and thus called "conventional crude oil." On the other hand, heavy and extra heavy crudes are mainly located in Mexico and Venezuela, and bitumen in Canada. These crudes are recovered by unconventional methods and hence are called "unconventional crude oil."

The next most important characteristic of crude oil is based on the boiling point distribution of the hydrocarbons. The very first step for crude oil when it enters the refinery from a pipeline is distillation. It undergoes atmospheric distillation followed by vacuum distillation in order to recover various fractions of hydrocarbons of different boiling point ranges in a crude oil. The relative proportions of the hydrocarbon fractions present in a crude oil may affect the downstream refining processes. Details of the commercial distillation process is beyond the scope of this book; however, details are available elsewhere[1].

However, a quick distillation analysis of the composition of hydrocarbons is usually done using gas chromatographic techniques before the actual distillation in order to anticipate the composition of various fractions. A typical boiling point distribution of a crude oil is usually represented by a distillation curve or simply by a simulated distillation (SimDist) curve. The curve represents the cumulative amount of the fraction that is distilled (either in volume or in weight) against the corresponding temperature of distillation.

The average density of crude oil processed by US refineries falls within 32 ± 1 °API. A typical SimDist curve of a crude oil of API 32 is shown in Fig 1.3.1. It is to be noted here that the crude oil analyzed by the author may not be a crude from a single region; it could be a blend of various crudes with blend API 32, made in order to process in a refinery. Thus, the SimDist of another crude of API 32 may differ from this curve.

SimDist of a crude oil by GC method is done according to the ASTM method D-2887. When crude oil falls into the category of extra heavy or bitumen, which contains a large quantity of undistilled residue, the SimDist is done using ASTM method D-5307. This method is capable of measuring the undistilled portion through back calculation using some internal standard. Residue, by definition, consists of the hydrocarbons not distilled under vacuum having boiling point above 525 °C or 1,000 °F.

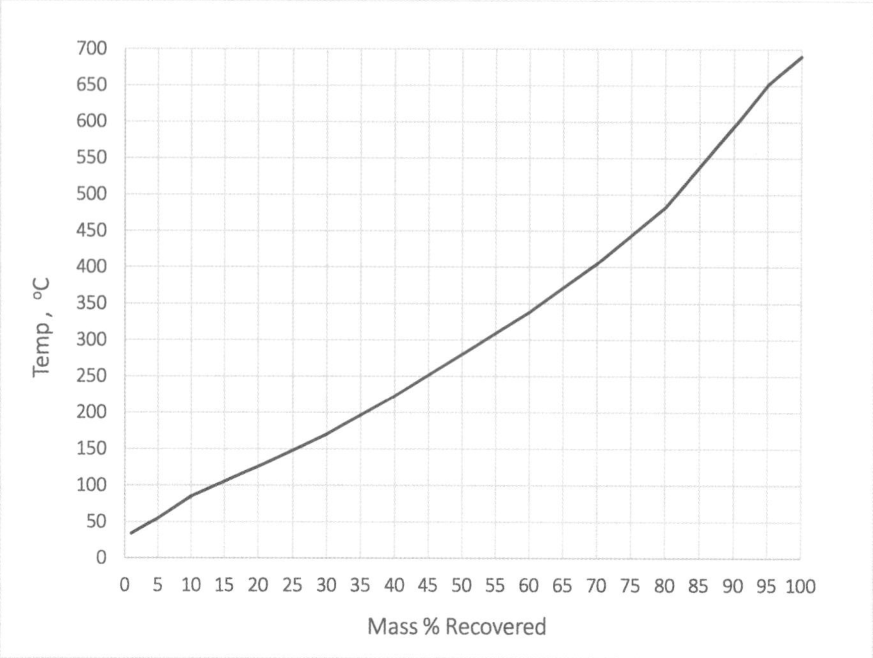

Fig. 1.3.1. Simulated Distillation curve of a typical conventional crude oil of API 32

1.4 Heavy Oil / Bitumen

Beyond C_{40} range, hydrocarbons are usually non-volatile at normal atmospheric conditions and mostly have a polycondensed aromatic structure. They are present in quite large quantities in the extra heavy oil or bitumen range crude, which are solid at atmospheric temperature. The heavy oil in the western Canadian region of Alberta is called bitumen. Though the API gravity is almost the same (<10) as that of the extra heavy oil (Table 1.3.1), it is highly viscous as compared to the heavy oils found in other parts of the world. Bitumen contains a significant amount (> 50 wt%) of material that is not distillable even under commercial vacuum unit.

In the heavier crude oil, the condensed aromatic and naphthenic structure becomes more complex with the replacement of C- atoms with the heteroatoms such as sulfur, nitrogen and oxygen atoms. The physical and chemical properties of these compounds with heteroatoms are quite different from the parent hydrocarbons from which they are derived. There are also molecules where metals are attached in the cyclic structure as organometallic compounds known as "porphyrin." The most prominent metals are nickel and vanadium in the heavy crude oil. These heavy hydrocarbons are a complex mixture of poly-condensed aromatic structure with metals and heteroatoms notorious for bad behavior called "asphaltenes" (for details please see elsewhere[2]). As discussed earlier, crude oil is characterized by the boiling point ranges or distillation curve, but heavy oil / bitumen consists of non-volatile resid and hence cannot be characterized fully by distillation curve only. Thus, it is characterized based on solubility of four major fractions, well known as SARA analysis, utilizing different solvents and adsorbing material in a column chromatography as shown in Table 1.4.

Table 1.4 Classification of heavy crude / bitumen in various fractions

Fraction	Solvent type / column chromatography material
Saturates	Soluble in n-paraffis (C5 or C7) / elutes through silica alumina column
Aromatics	Soluble in n-paraffis (C5 or C7) / adsorbes in silica alumina column – elutes with paraffin – toluene mixture
Resins	Soluble in n-paraffins – adsorbes in silica gel column-elutes with toluene
Asphaltenes	Insoluble in n-paraffins – soluble in toluene
Inorganic / coke	Insoluble in toluene

When this heavy crude oil is subjected to thermal processing, every fraction behaves differently as they have different cracking behaviors. Saturates are paraffinic hydrocarbons, and they crack more easily than the other fractions. Although all other fractions contain aromatic hydrocarbons, the number of rings condensed together increases from aromatics to resins to asphaltenes. The aromatic fraction consists of simple aromatics bonded together with carbon chains; they break at the weakest C-C bonds, and aromatics go to the lighter fractions after thermal cracking.

Resins and asphaltenes contain heteroatoms (sulfur / nitrogen / oxygen) in their ring structure, and they take part in the thermal processes. Asphaltenes also consist of metals as "porphyrin" structure. Asphaltenes are the core structures-type molecules that act as a precursor of coke formation—and mostly get converted into solid carbon material known as "coke," discussed in detail in Chapter 7.1.

So readers will understand the complex nature of the various molecules or groups present in the resid fraction or bitumen itself, a hypothetical representation of the structure in simple format is shown in Figure 1.4. The core innermost structure is the asphaltenes made of a poly-condensed aromatic nucleus. Its ring structure also contains sulfur, nitrogen, or metals, as shown with the letters S, N, and M. The next layer consists of resins; they are very similar to the asphaltenes but of lower molecular weights, i.e., they contain loosely bonded rings and heteroatoms. The outer layer contains aromatics and saturates. They are also part of the extended branch chains attached to the core structure of asphaltenes and resins. The heteroatoms in the outer layers are represented by "x" attached to the long chain of hydrocarbons, where "R" represents the chain length or the number of carbon atoms present per molecule.

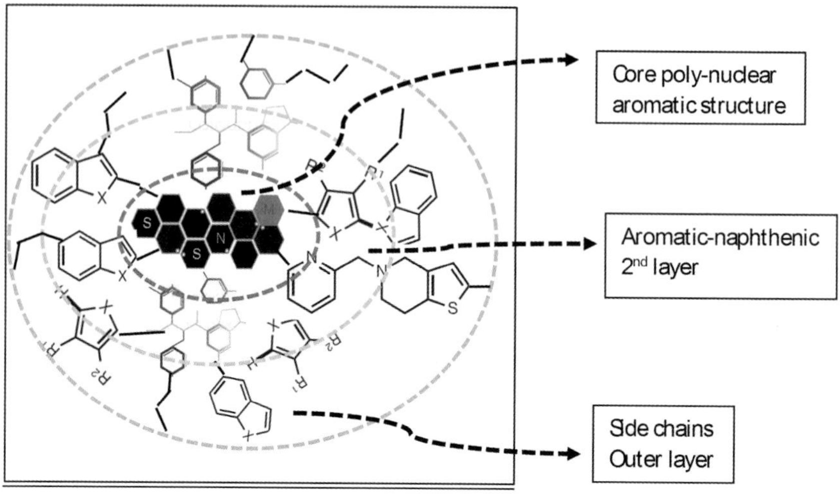

Fig. 1.4. Plausible molecular orientations of a typical bitumen

The number of hydrogen atoms per molecule are in the following order:

saturates > aromatics > resins > asphaltenes

In other words, the H/C atomic ratio decreases from saturates to asphaltenes, which also represents the aromaticity of a fraction. The higher the H/C ratio, the better the quality of the fraction of that crude oil. Thus, the value of the heavy crude oil decreases with the increase in asphaltene content.

Another factor that plays an important role in thermal processing of heavy oil or bitumen is known as Conradson carbon residue (CCR) or micro carbon residue (MCR). This is an indication of the coke-forming tendency during the thermal process. Asphaltenes usually have the highest CCR / MCR values, or the highest tendency to coke formation.

1.5 Types of reservoirs

Reservoirs are made of shale: a combination of silica / clay / calcite. Shale is a rock formed by deposit of the sediments from marine water, mostly made of clay and silt. Different combinations of those materials make different qualities of shale, e.g., carbonate makes the shale porous.

Shales are 400–500 million years old (before the dinosaurs' age). Age, depth, temperature, and pressure of the shale change the thermal maturity of the hydrocarbons. That makes different qualities of shale in the same region. As the maturity increases, the quality of the hydrocarbon distribution changes as follows:

$$\text{Liquid oil} < \text{Wet-gas} < \text{Dry-gas}$$

In other words, the molecular size of the hydrocarbon decreases as the maturity of the reservoir increases.

The amount of hydrocarbon in the shale formation also varies, as measured by Total Organic Content (TOC). If the TOC is less than 2%, it is considered not a commercially viable reservoir.

As shown in Figure 1.5, the quality of the reservoirs varies with the types of shale deposit with time. The permeability of the shale for clay-type material is almost zero (<0.01 mD, milliDarcy), and the quality of the reservoirs is extremely poor, which is a solid rock. The permeability increases between 0.01 to 0.1 mD for calcite / dolomite-type material, and further increases between 0.1 to 1.0 mD for rocks of limestone quality. Reservoirs with permeability greater than 1.0 mD are considered good quality reservoirs of high porosities. All conventional crude oils are usually found in this type of reservoir.

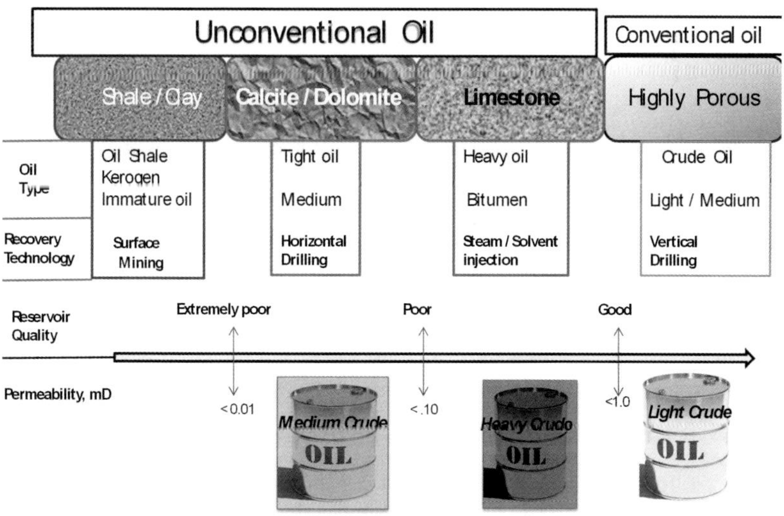

Fig.1.5. The nature of shale deposits in various types of reservoirs

The types of hydrocarbons and the qualities of those hydrocarbons are different for different types of deposits, and because it is trapped such that it is difficult to get it out, they need different technologies to recover the oil or hydrocarbons.

The oil present in the poorest quality clay reservoir is called oil shale (not to be confused with shale oil / tight oil)—actually there is no trapped oil here. The organic matter present is kerogen and is part of the shale itself close to the surface. Kerogen is not even an oil; the shale, which is not even a shale, is another misnomer in the oil industry. The shale is first recovered through surface mining, and the organic material kerogen can be separated only by a thermal treatment known as "retorting."

The next matured reservoir is mostly made of calcite / dolomite. In this type of reservoir, the hydrocarbons are trapped inside the rocks so tightly that they cannot escape by themselves. So the only way the hydrocarbons can escape or recover from the rocks is by artificially cracking the rocks and opening up the pores. This opening up of the pores is done with the help of a new technology, horizontal drilling, in combination with hydraulic fracturing, widely known simply as fracking (discussed in detail in Chapter 5). Though it is difficult to recover the oil, the quality of the oil is relatively good and equivalent to a medium-grade crude.

The next in the series is a better-quality reservoir mostly made of limestone. This is of higher porosity with the permeability ranging between 0.1 to 1.0 mD (milli Darcy). However, the oil in those reservoirs is extremely poor with very high density (API < 12) and falls in the category of heavy oil and bitumen. The difficulty in the recovery or extraction of those heavy hydrocarbons, especially bitumen, led to the development of a new technology, widely known as "SAGD," that involves addition of high-pressure steam or solvent into the reservoir (see reference 2 for detail).

The best quality reservoirs with very high permeability (> 1.0 mD) are the major sources of oil in the world. The reservoir is highly porous, and hence the hydrocarbon molecules are mobile inside the porous rocks. There is a natural seal or cap rock over that oil zone that prevents escape of the hydrocarbons to the surface. The oil and gas can be recovered by the simple conventional method of vertical drilling. It's estimated that over a million wells have been drilled since the first well in the late 1940s (vertical drilling). The conventional crudes are of much better quality as compared to the heavy crudes.

1.6 Definitions

- Conventional Oil
    - When a well is drilled into a reservoir, the oil characteristics permit it to flow back to the surface by itself.
- Unconventional Oil
    - The oil cannot be recovered by conventional means either because
        - the porosity and permeability of the reservoir do not permit the flow of the oil

          or

    - the density and viscosity of the oil is high enough that it cannot be produced or transported by itself.

- Porosity
    - Porosity is the capacity of a rock to store fluid.
    - It is a ratio of open (or porous) volume to the bulk volume.
    - Different grain size and packing arrangements will result in different porosity values.
    - Individual pore spaces decrease in size with decreasing grain size.

Figures 1.6.1 and 1.6.2 show how the porosity varies with the size of the grains in the reservoirs. Bigger-sized grains create higher porosity and smaller-sized grains create tight packing, hence decreasing the porosity.

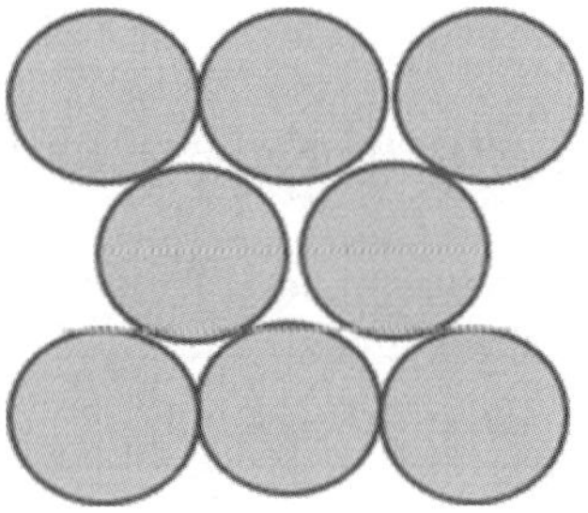

Fig. 1.6.1. Bigger grain size creates bigger pore volume, hence higher porosity.

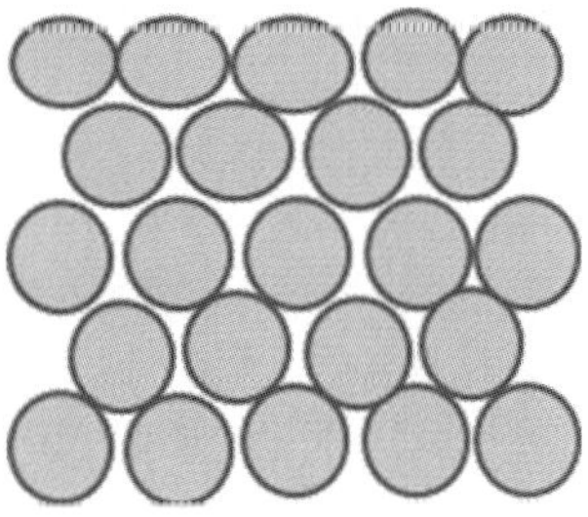

Fig. 1.6.2. Smaller grain size creates tight packing with smaller pore volume and less porosity.

- Permeability
    - Permeability is a measure of the capacity of a fluid to flow through rock.
    - The rock must have connectivity of the pores.
    - High permeability often goes hand in hand with porosity.

Figure 1.6.3 illustrates how the connectivity of the pores allows the flow of oil and hence increases the permeability of the reservoir.

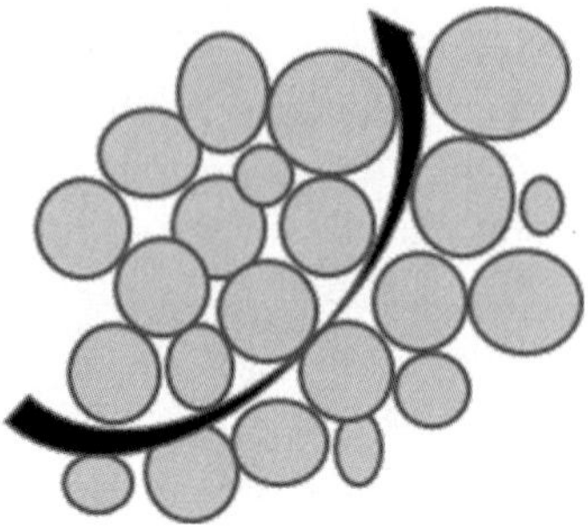

Fig. 1.6.3. Oil flows in the direction of where open connected pores are available.

1.7 Carbon-type distribution in crude oil

A conventional light crude oil contains hydrocarbons with carbon number starting from five carbons (C5) per molecule up to 50 carbons (C50) per molecule. Only heavy crude oil or bitumen may contain paraffinic hydrocarbons beyond 50 carbons or polycondensed aromatics of lower carbon number but boiling point higher than that of the C50 paraffinic hydrocarbon.

Figure 1.7 illustrates equivalent carbon number distribution in a typical crude oil of API 32, derived from the simulated distillation curve of the oil. The groups containing a selective range of carbon numbers are basically divided based on the boiling points of the corresponding paraffinic carbon, as shown in Table 1.7.

Results show that in a typical conventional light crude oil processed in United States refineries, the concentration of hydrocarbons decreases with increasing molecular weight from naphtha (C5–C10) up to C50 range hydrocarbons. Interestingly, concentrations of naphtha, kerosene, and diesel decrease rapidly from 30 wt% to 20 wt% to 10 wt%, respectively. More than half of crude oil consists of hydrocarbons below C20 and may contain some aromatics.

Figure 1.7 also shows that about 10% of crude oil is made of resid fraction (1050 F+), which consists of un-distillable C50+ range combined material. It is to be cautioned here that in the resid fraction, most of the materials are highly poly-nuclear aromatic in nature and hence their carbon numbers per molecule may be less than C40, but their boiling point is higher than equivalent C50 paraffinic hydrocarbon. Also, it is to be noted here that the composition changes for similar light crude oil from region to region.

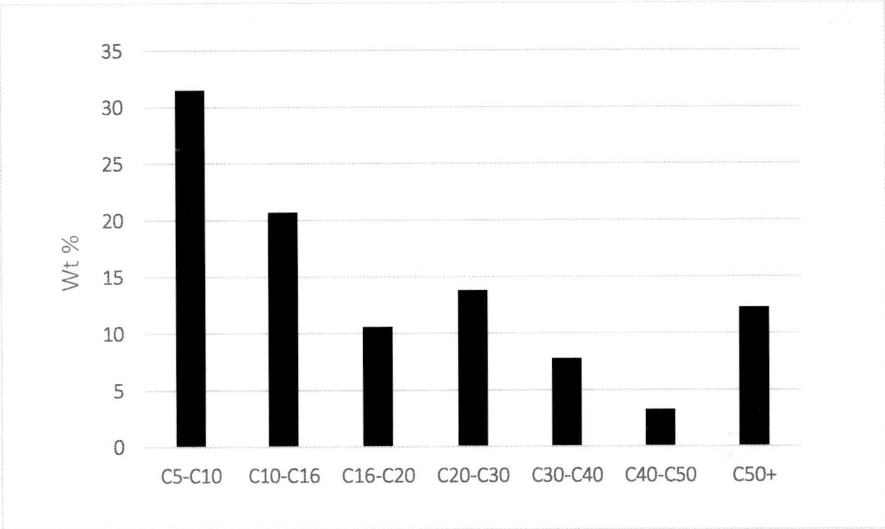

Fig. 1.7. Carbon number distribution in a typical crude oil of API 32

Table 1.7. Various groups based on carbon number distribution with corresponding boiling ranges

Carbon number range	Group name	Temp range, °F	Temp range, °C
C5 – C10	Naphtha	IBP - 350	IBP - 175
C10 – C16	Kerosene	350 - 550	175 - 290
C16 – C20	Diesel	550 - 650	290 - 345
C20 - C30	Light gas oil	650 - 850	345 - 455
C30 - C40	Heavy gas oil	850 – 975	455 - 525
C40 - C50	Fuel oil	975 - 1050	525 - 565
C50 +	Resid	1050 F+	565 C +

1.7.1 Paraffins

Paraffins, also known as alkanes, are the backbone of the hydrocarbon structure in crude oil. The majority of the conventional light crude oil is paraffinic in nature, i.e., consists of more than 70–75% paraffins. It is made up of a homologues series represented by the empirical formula C_nH_{2n+2}, where all the carbon bonds are saturated with hydrogen atoms. The series starts and ends with a CH_3- group and is attached in between with bridges of (– CH_2 –)n groups. When the groups are present in a straight chain, they are named as normal- paraffins and when they are present as a side chain, they are called iso-paraffins. The symbol 'n' in the series represents the number of C-atoms in a molecule and is always a positive integer number. In a conventional light crude oil, the length of the carbon chain may extend up to C40 or C50 carbon numbers. However, except for the first three hydrocarbons in the series, C_1 to C_3, as the carbon number increases, the straight chain hydrocarbons branch from the skeleton structure, and quickly the probability and possibilities of the number of branches increase as the molecular weight of the hydrocarbons increases.

For the sake of simplicity, an example is illustrated below showing the number of possible isomers for the simplest case of C_6-paraffin or hexane:

Here the symbol

represents a carbon atom with its four covalent bonds.

(i) Thus, n-hexane is shown here as

(ii) 2-Methylpentane or iso-hexane

(iii) 3-Methylpentane or iso-hexane

(iv) 2,3 Di-methylbutane

(v) 2,2 Di-methylbutane

In a normal series of hydrocarbons each and every –C-C- bond contributes its energy almost equally, and the properties in the series increase incrementally; however, in case of isomers, the contribution of bond energy is apparently different depending on the position.

Paraffins also exist in the form of cyclic structures cycloparaffins, also known in the oil industry as "naphthenes," which can be classified as saturated aromatics, not to be confused with naphthalene, which is an aromatic hydrocarbon.

Interestingly enough, the molecular shape of the cycloparaffins changes drastically from C_3 to C_6 cyclic structure, as shown below:

(i) Cyclic C_3 – cyclo-propane is triangular in shape

(ii) Cyclic C_4 – cyclo-butane is represented by a square

(iii) Cyclic C_5 – cyclo-pentane is a pentagon

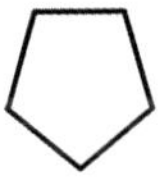

(iv) Cyclic C_6 – cyclo-hexane is hexagonal in structure

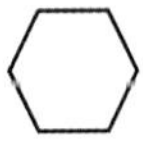

The C-C bond energies increase from C_3 to C_6 cyclo hydrocarbons; thus, the thermal stability of the cycloparaffins increases with the increase in carbon number. When the cyclic compounds are subject to thermal processing, they have two options:

either (i) break the C-C bond

or (ii) dehydrogenate by releasing hydrogen

Through dehydrogenation, it results in corresponding double-bond hydrocarbons such as cyclo-propene, cyclo-butene, cyclo-pentene, and cyclo-hexene.

1.7.2 Olefins

As discussed above, the paraffins are saturated hydrocarbons; on the other hand, when the hydrogen atoms are missing from the skeleton, they are named as "unsaturated hydrocarbons" or technically referred to as "olefins" or also known as "alkenes" represented by the (– C = C –) moiety with an empirical formula of C_nH_{2n-2}. The number of double-bond isomers increases with the increase in molecular weight. Molecules with two or three double bonds are known as "dienes" or "trienes," respectively.

The third series of unsaturated hydrocarbons with triple-bond moiety such as C ≡ C are known as "alkynes." It is to be noted here that a small variation of double or triple bonds in a molecule brings a significant change in the reactivity of the hydrocarbon. Conventional crude oils do not contain any olefins.

1.7.3 Aromatics

The six-member unsaturated cyclic hydrocarbons are known as aromatics, and the lowest member of the aromatic series is benzene, which is fully unsaturated cyclohexane with the loss of six hydrogen atoms. Aromatics are found in crude oil but mostly attached to the alkyl side chain. Substituted aromatics of various molecular weights are found throughout the boiling range of crude oil. Heavier crude oil contains polycondensed aromatic ring molecules.

Benzene is known to have a highly thermally stable structure, greater than its usual double-bond energy of 80 kcal/mol. The reason being three double bonds in benzene have a resonating structure that provides additional thermal stability to the benzene ring. The structure was proposed by the German chemist Kekule (see chart in the Preface), who claimed to have dreamed the structure and proposed it only one and a half century ago.

The benzene ring double-bond resonating structure is usually represented as follows:

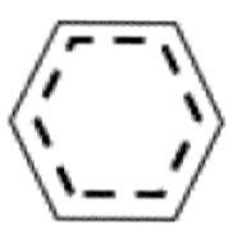

The H- atom attached to any of the six C-atoms in the benzene ring could be further replaced by other series of alkanes or alkenes, etc., the lowest member of the series attached to a methyl group being toluene and the next attached to an ethyl group is ethylbenzene and so on. The aromatic structure gets more and more complicated as the molecular weight of the hydrocarbons increases.

Isomers of ethylbenzene, where two methyl groups are attached to two different C-atoms of the benzene ring, are named xylene, and there are three possible xylene isomers: ortho, meta, and para, depending on the positions of the methyl group as shown below:

Ethyl benzene

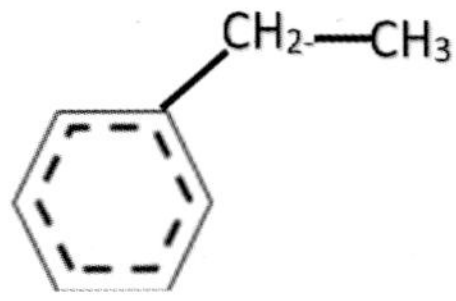

Ortho -Xylene

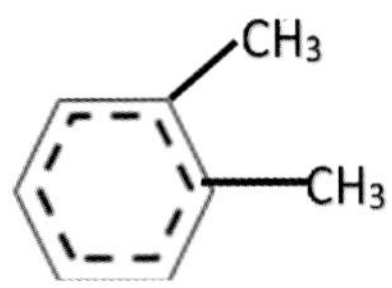

Meta-Xylene

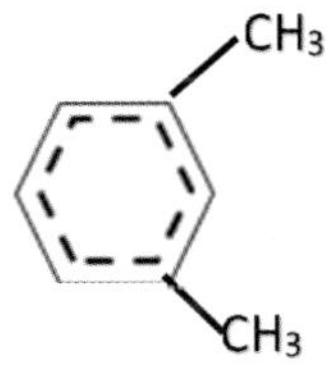

Para-Xylene

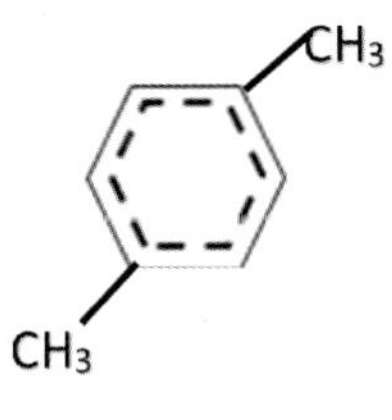

There are also fused cyclic structures such as two-fused cyclohexane, which is called decalin $C_{10}H_{16}$, and the corresponding fully unsaturated hydrocarbon is named naphthalene $C_{10}H_8$:

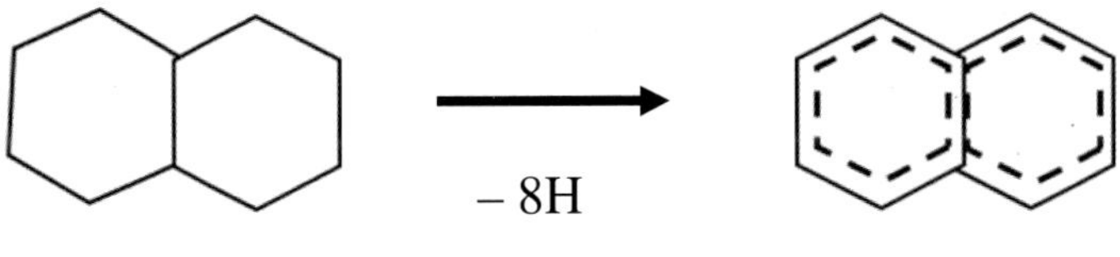

Decalin Naphthalene

And the next in the series with a three fused aromatic ring structure, $C_{14}H_{10}$, is named anthracene and its isomers is called phenanthrene.

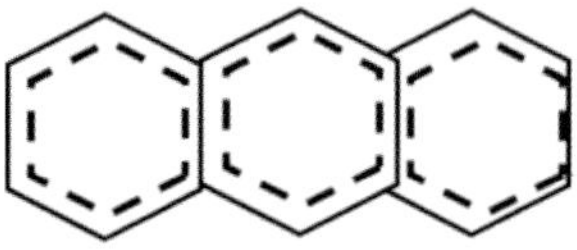

Anthracene

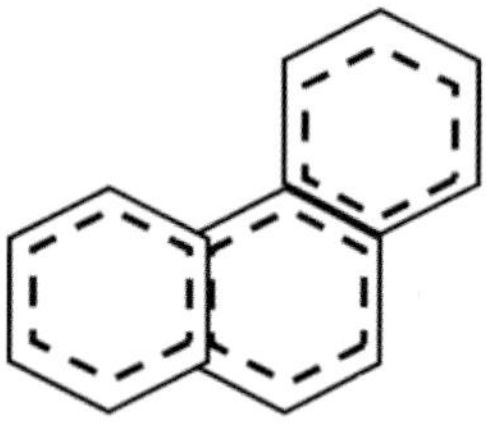

Phenanthrene

The series goes on with multiple-ring hydrocarbons fused together being partially saturated and partially aromatic in nature. The lowest member in this series is tetralin, a condensed structure of cyclohexane and benzene. The complexity of the poly-condensed aromatic structure increases with the increase in molecular weight of the petroleum hydrocarbons.

1.8 Crude oil marketing

There are different crude oil markers, each one representing crude oil from a particular part of the globe to a specific market. There are three major markers used in pricing of crude oil across the globe:

(i) WTI (West Texas Intermediate) for the US market

(ii) Brent crude marketed from London for the European/West African market

(iii) Oman / Dubai for crude oil grades in the Persian Gulf for the Asian market

Properties of those three marker crudes are shown in Table 1.8.1. Brent and WTI are quite close to each other, and both are low sulfur sweet crude. Dubai and Oman crudes are heavier and sour crude with sulfur concentration greater than 1%. All crudes have quite low acid numbers as per crude specifications.

Recently, with increasing volume of heavy oil development in Canada, they have introduced a special heavy oil marker called Western Canadian Select (WCS) and another for the upgraded synthetic crude called Syncrude Sweet Blend (SSB), closer to the light crude marker.

Table 1.8.1. Major characteristics of worldwide marker crudes

Marker crude type	Brent	WTI	Dubai	Oman
Properties				
°API	38	40	31	30
S , wt%	0.45	0.35	1.54	1.45
Acid number, mg KOH/gram	0.01	0.09	0.05	0.61

Also, with increasing oil export from Russia, they also introduced their own marker blend, REBCO, whose properties are shown in Table 1.8.2, along with the other marker crudes.

Table 1.8.2. Characteristics of Canadian and Russian marker crudes with respect to WTI and Brent crude

Properties	API	Density, Kg/m^3	S, wt%	CCR, wt%
SSB, Canada	31-33	860-870	0.1–0.2	<0.1
WCS, Canada	19-22	920-940	2.8-3.2	9-10
Russian Export Blend (REBCO)	31.5	870	1.2	3.4
Brent crude, North Sea	38.0	833	0.45	2.13
WTI, USA	40.8	821	0.35	1.08

Brent crude is priced every day from London, and it covers almost 2/3 of the world market including Asia, Europe and Africa. Brent crude is more widely used then the other two Asian crudes from Oman or Dubai. As shown in Table 1.8.1, Brent is a lighter and sweeter crude than those from Oman and Dubai. Brent crude is actually a mixture of four North Sea crudes, and being offshore, it is easy to transport through the water line. Almost all other crudes in the world are priced based on the properties of Brent crude.

The bench mark crude in the US is called West Texas Intermediate (WTI), which is quite similar to Brent crude, a light sweet crude. Usually WTI tracks the price of Brent in the market, but is always priced a little lower than the Brent, mostly because of the distance between America and Europe. Though WTI is named after Texas, it is priced daily in Cushing, Oklahoma, which is the hub of the oil storage depot. Almost all major pipelines in the US meet at Cushing and are then directed towards the Gulf Coast.

Western Canadian Select (WCS) is a recently created bench mark crude from Canada. It was specially created in order to match the unconventional Canadian heavy crude / bitumen with the world market through WTI. WCS is an upgraded heavy crude oil, blended with several other upgraded crudes and conventional crudes. Canada also introduced another marker crude in order to match the world sweet crude market. It is a highly upgraded sweet synthetic crude called SSB. As shown in Table 1.8.2, the properties of WCS and SSB are assigned within a short range and not to a specific property, as is the case for the other benchmark crudes. Details of the Canadian crudes are discussed elsewhere[2].

Properties of the benchmark crude oils (Figure 1.8.1) indicate that the Russian crude REBCO has the highest amount of resid; hence, the primary market should be the refineries with a large capacity of delayed coker plants. On the other hand, WCS has the highest amount of gas oil-range hydrocarbons, indicating that a refinery with a larger capacity FCC plant will attract this crude. Brent crude has the highest amount of distillates, which is valuable for diesel production, whereas WTI and Brent both consist of large amounts of naphtha, which is good for the refineries producing gasoline for the auto-transportation market.

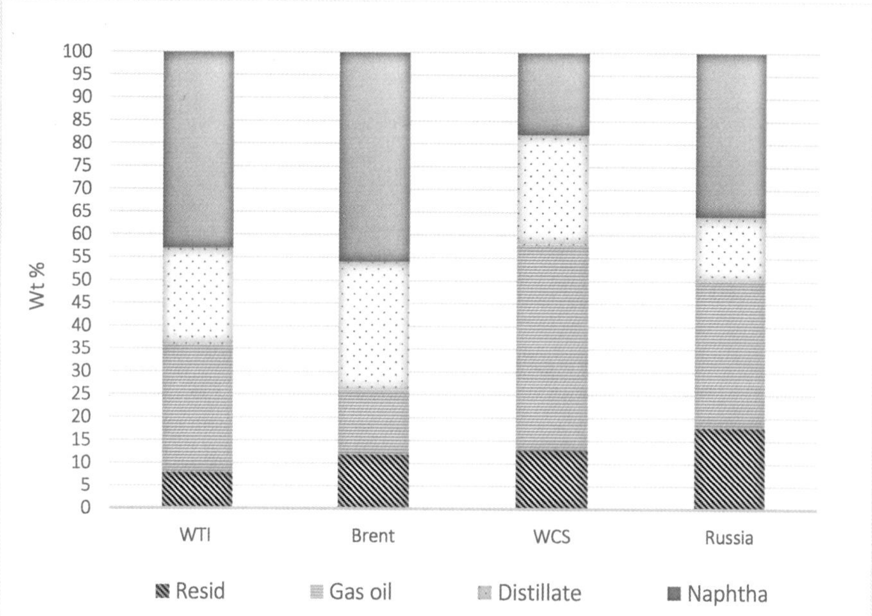

Fig. 1.8.1. Worldwide benchmark crudes compositions (resid: bottom, and naphtha: top of the bar graph)

1.9 Crude oil properties

Thus, not all crudes are attractive to all refineries; it depends on its location and their product market demand. Volatility in the market creates uncertainty that has a significant impact on the refinery operation. Also the market is continuously shifting mostly due to environmental reasons such as ultra-low-sulfur gasoline, eliminating diesel, introducing bio-fuels, IMO2020 low-sulfur bunker fuel, increasing demand for petrochemicals, etc. Those are serious processing challenges for the refiners.

A refinery will purchase a crude oil strictly based on its properties because a refinery can handle a crude oil within a certain specification only. For example, when a crude oil enters the gate of a refinery, it is first fed to an atmospheric distillation unit followed by a vacuum distillation plant. The plant is designed to handle crude oil within a certain API range only; for the US refineries it is around 32 API:

- API gravity provides a number to convert mass into volume and vice versa. A refinery then estimates the amount of crude it will be processing through which plant:
  - if the crude is very light (>40 API), the atmospheric distillation column overhead section becomes limiting
  - alternatively, if the crude is heavier (<15 API), the bottom of the vacuum distillation unit may become limiting
  - if the crude is heavier the asphaltenes content can cause fouling in the furnace coil operation
- Properties like viscosity provide an indication of the mobility of crude oil, which is influenced by the temperature.
- Pour point / Cloud point indicates the relative amount of wax present in crude oil (especially important for the new light oil in the market).
- Metals content (usually Fe, Ca, Mg, Ni, V, etc.) can cause problems for the desalter operation in a refinery and also affects the downstream catalyst activities.
- Crude oil with an acid number (TAN number, mg KOH /g crude) beyond an acceptable limit (usually >1.0) cannot be accepted due to corrosion issues.
- Carbon residue number (especially important for heavy crudes) indicates the tendency for coke formation.

Sulfur content in crude is an important factor, an indication of how much secondary processing is required to meet the target finished product qualities. The properties of primary processed product can limit the secondary processing units as well as the final product specification. As such, known industry data can be used to predict potential risks for the new crude. The compatibility of new crudes with crudes already being processed must be considered.

References for Chapter 1

1. A. Nag, Distillation and Hydrocarbon Processing Practices, Publisher PennWell Corporation, USA (2016)
2. Dwijen Banerjee, *Oil sands heavy oil bitumen, from recovery to refiner*, Publisher PennWell Corporation, USA (2012)

2

Fundamentals of thermal processing

2.1 Background

Development in science is quite painful and time-consuming. Most of the major discoveries in the world were made in the last couple of centuries. Just over two centuries ago, Benjamin Franklin discovered what he called "charged fire," which was actually electricity, not known at that time. During the same period, we actually did not know even the definition of "heat"; scientists considered this as "flow" of some unknown matter. Only in the mid-nineteenth century was heat defined as an energy, and one of the most revolutionary applications was the discovery of the "steam engine." Then, the science of energy was developed, and scientists explained that heat is a "flowing energy" that could be converted into another form of energy, but cannot be destroyed. Apparently, this was called "Thermodynamics," and the "First Law of Thermodynamics" was born in England, where the subject of "energy" was pioneered at that time. Then, the study moved to France where the science of energy evolved further with the discovery of the "Second Law of Thermodynamics," which explains that heat can move only in one direction, defined as "Entropy."

The scope of this book is to summarize the basic information available from the pioneer workers in this field of thermal processing of hydrocarbons. It is true that since then, a large volume of work has been carried out with complex mixtures of hydrocarbons by hundreds of researchers around the world, and thousands of papers have been published in various journals. However, in order to keep the book concise, and assuming the readers have some knowledge in this area, the author concentrated mostly on the world-famous pioneer workers in this field and presented the fundamental theories and their industrial applications in simple language for the easy understanding of the readers. For example, the reaction mechanisms of thermal cracking of hydrocarbons have been studied extensively; however, the debate continues, as the reactions are so complex that the analyses of intermediate products are very difficult in order to predict the reaction mechanisms accurately. The "free radical" reaction mechanisms and the corresponding kinetic data are still valid and could be applied for commercial use, in spite of the fact that even after half a century, still there is debate going on between the "free radical" vs. "molecular" and "heterogeneous" vs."homogeneous" reaction mechanisms. Because of that, the author is not inclined to support all of the work reported in various publications, and has therefore omitted individual publications.

Some of the most prominent scientists of early days were Hinshelwood[1], Eyring[2], Pauling[3], Polanyi[4], Rice[5], Steacie[6], Dickenson[7], and Laidler[8] who gave considerable attention to this area of research. The pioneer workers Steacie and Rice gave us the concept of the reaction mechanism of thermal cracking of hydrocarbons based on "free radical formation" in the early 1950s. According to the theory, in the presence of excess energy, a paraffinic hydrocarbon molecule breaks the weakest C-C bond and becomes a free radical. The free radical itself then colloids with another molecule and extracts a hydrogen atom, and it becomes a free radical itself. The chain reaction of the formation of free radicals continues. Once the free radical is formed, then it has either the possibility of further breaking down, most preferably the C-C bond in the β position of the radical, or dehydrogenating to form olefins. The reactions are initiated by the direct application of heat with the formation of free radicals. The half-life of the free radicals is a matter of fractions of a second. The chain reactions that follow are extremely rapid. Measurement of the intermediate products needs a very special spectroscopic instrument.

2.2 Reaction mechanisms

A basic simple reaction mechanism of the heat treatment of a compound AB is shown as follows:

$$AB \quad \leftrightarrow \quad A + B \qquad \text{.........(2.2.1)}$$

where the compound AB breaks down into two molecules A and B. The reverse reaction indicates that only a portion of it breaks down, and they can recombine, and at equilibrium, a portion of the total mass remains in the broken stages of A and B. The equilibrium may shift depending on the reaction conditions. In a catalytic process, the same compound may break down into "ions" having positive and negative charges. In a solution, almost all reactions are "ionic"; they are "reversible," and they reach an equilibrium.

When any organic molecul, big or small, is subject to heat, based on the actual conditions, the molecules are shifting continuously between various arrangements depending on their energy levels and surroundings. Smaller molecules move at a faster speed as compared to the bigger molecules in a continuous reactor system. Thus, the smaller molecules have shorter residence times than the bigger molecules under the same energy conditions, and they undergo different degrees of reactions. There is actually no gravitational force in effect while hydrocarbons are under thermal processing. The bigger the molecule is, the longer the time it sits there; it breaks into several pieces in the form of free radicals. These free radicals wait with their eyes open to bind together with any other surrounding organic molecules it can see. These reactions are known as "thermal cracking," as the hydrocarbons break into various sizes of fragments. The cracking process is an "endothermic" reaction, which needs excess heat to takes place, i.e., higher temperatures. The process takes place almost randomly. The amount of heat required for cracking a molecule depends on the bond energy, or in other words, the energy required to break the bond. Organic hydrocarbons are mostly made of covalent bonds, and their bond energies vary depending on the carbon-carbon and carbon-hydrogen chain length and types of carbon, such as paraffinic, olefinic, or aromatic, etc. In petroleum, the carbon is also attached to other types of elements such as sulfur, nitrogen, and oxygen. The bond energy varies depending on their positions and to which elements it is attached.

It is to be noted here that the thermal reactions of hydrocarbons are in fact a very complex system. Simultaneously, there are several consecutive and parallel reactions taking place which could be explained based on molecular or radicalic reaction mechanisms. The possible reaction mechanism, however, depends on how one can arrive at the formation of various product distributions based on the thermodynamic data.

Once the appropriate energy is set up in a reactor and the hydrocarbons are fed, the molecules will break in sequence depending on the bond energy of that particular bond, going from the weakest bond to the strongest one. In practice, sufficient energy is always added into a reactor to break most of the desired molecules under investigation. A different state of molecules has a different molecular energy level, and they could flip between different structures such as primary to secondary to tertiary. The stability of the molecules depends on their energy levels.

Although the detailed mechanisms are not well understood, let us concentrate on simple thermal cracking processes. When a hydrocarbon is subject to a thermal process, there is rupture of the C-C bond at the weakest point under the reaction conditions and it forms "free radicals"—let us represent it by the symbol (*) as follows:

$$C_{n1} - C_{n2} \longrightarrow C_{n1}^{*} + C_{n2}^{*} \qquad \text{......(2.2.2)}$$

where n represents the carbon chain length.

The free radicals are so reactive that the half-life may vary from milliseconds to seconds. In case of thermal cracking reactions of hydrocarbons, almost all reactions are radicalic and "irreversible." However, all catalytic reactions follow ionic pathways, and they could be reversible.

According to the works of the pioneer scientists, the fundamentals of thermal cracking reaction mechanisms could be summarized as follows:

Consider that a typical structure of hydrocarbon consists of aromatic (Ar_n) and aliphatic (R_n) groups as follows:

$$Ar_1 - R_1 - C\text{-}\ C - Ar_2$$

When a complex molecule such as shown above is thermally cracked, it undergoes three basic radicalic reaction pathways:

1st step—Initiation reactions

Scission of intramolecular bonds, and formation of free radicals:

$Ar_1 - R_1 - C\text{-}\ C - Ar_2 \longrightarrow Ar_1 - R_1 - C^* + {}^*C - Ar_2$

(breaking of C-C bond).....**(2.2.3)**

$Ar_1 - R_1 - H \longrightarrow Ar_1 - R_1^* + H^*$

(breaking of C-H bond).......**(2.2.4)**

2nd step—Propagation reactions

There are several options for simultaneous reactions of free radicals:

$Ar_1 - R_1 - C\text{-}\ C^* \longrightarrow Ar_1 - R_1^* + C_2H_4$

(breaking at β bond) .... **(2.2.5)**

$Ar_1 - R_1 - C^* \longrightarrow Ar_1 - R_1 = C + H^*$

(dehydrogenation) .... **(2.2.6)**

$Ar_1\text{-}H + R^* \longrightarrow Ar_1^* + RH$

(abstraction of H_2) .... **(2.2.7)**

$R_n^*\ (n>6) \longrightarrow$ 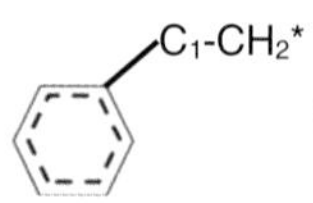

(cyclization) .... **(2.2.8)**

3rd step—Termination reactions

Stabilization of the free radicals through recombination:

$Ar^* + Ar^* \longrightarrow Ar\text{-}Ar$ ....**(2.2.9)**

$R_n^* + Ar^* \longrightarrow R_n\text{-}Ar$ .... **(2.2.10)**

$R_1^* + R_2^* \longrightarrow R_1\text{-}R_2$ .... **(2.2.11)**

It is essential to point out here that in a system with free radicals, such as in a thermal process, propagation steps (described in step 2 above) are the most important steps that proceed in a chain mechanism, which is responsible for various product distributions and where the transfer of energy takes place.

On the other hand, when two free radicals recombine, as shown above in step 3 termination reactions, very little energy transfer takes place.

Shown below is recombination of two radicals for the simplest case of methyl radicals:

$$CH_3^* + CH_3^* \longrightarrow CH_3 - CH_3 \qquad \text{.......} \textbf{(2.2.12)}$$

When two methyl radicals recombine, a dimer is formed, e.g., ethane; the energy transfer in the formation of dimer is negligible as compared to the other options available to the ethyl radicals as follows:

$$C_2H_5^* + C_2H_5^* \longrightarrow C_2H_4 + C_2H_6 \qquad \text{.......} \textbf{(2.2.13)}$$

where the ethyl radicals may undergo disproportionation reaction to form ethane and ethylene with some energy transfer.

The generally accepted most favorable theory of cracking of a radical is the β position to the free radicals in all cases. In case of alkyl aromatics, the most stable bond is the $C_{aromatic} - C_{aliphatic}$ bond, and hence the major product from thermal cracking of alkyl-aromatic is toluene and not benzene. The most stable radical known to have a longer half-life is a benzyl radical due to the resonance structure of the aromatic ring attached next to the radicals as shown below:

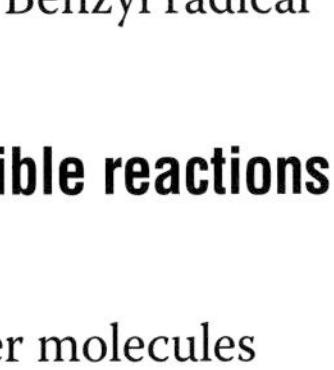

Benzyl radical

2.2.1 List of possible / plausible reactions of thermal processing of hydrocarbons:

(i) Cracking into smaller molecules
(ii) Dehydrogenation to olefins
(iii) Paraffin cyclization to naphthenes
(iv) Dehydrocyclization to aromatics (for >C6)
(v) Aromatization
(vi) Paraffins isomerization

(vii) Aromatic side chain cracking
(viii) Breaking of ring structure
(ix) Polymerization of aromatics
(x) Polymerization of olefins
(xi) Coking reaction—coke formation

2.3 Elementary kinetics

In considering the elementary thermal processing reaction of hydrocarbons, the most important questions that arise in this connection are the relationship between the heat supplied, which is directly related to the temperature of the system, and the rate of the reaction. The relationship proposed by the great Swedish scientist Arrhenius[9] (see the chart in the Preface) in the 1880s is still valid and well known as the Arrhenius Equation, shown below:

$$k = A \exp\left(\frac{-E}{RT}\right) \qquad \text{....... (2.3.1)}$$

$$\text{or } \ln(k) = \ln(A) - E/RT \qquad \text{....... (2.3.2)}$$

Where,
k = Rate constant
A = Pre-exponential factor
E = Activation energy
R = Gas constant
T = Absolute temperature, °K

The Arrhenius law is obeyed by almost all types of thermal reactions with significantly higher accuracy. It can be easily verified by plotting ln k against the reciprocal of absolute temperature 1/T, and always a linear relationship is obtained. The slope of the line gives the value of E/R so that the activation energy E can be calculated, using the gas constant value of R.

Rice and Steacie gave theoretical explanations based on free radical mechanisms as to how various products are formed from the thermal decomposition of a complex mixture of hydrocarbons. The most important conclusion that Rice and his coworkers reached was regarding the kinetic law. In spite of the complex behavior of the hydrocarbons, they presented arguments based on mostly simple "first-order kinetics." They explained that the formation of free radicals are most probably based on destroying the -C-C-linkage in the initiation steps (step 1, as explained in the previous section), and the "activation energy" of the first order kinetics is closer to the energy required to break the C-C bond, around 80 kcal/mol. However, they found that the experimental activation energy was lower than that, and they postulated that the overall kinetics is a combination of propagation and termination steps involved in the process, and those are not pure first-order reactions, which influences the overall activation energy.

In general, the important results that have been accepted by most researchers are that the overall reaction is of the first order over a wide range of conditions. They also confirmed that the first-order rate constant 'k_1' of the main thermal processing reaction is in the following order:

$$k_1 \text{ of } \textit{initiation} < k_1 \text{ of } \textit{propagation} < k_1 \text{ of } \textit{termination}$$

There had been a good debate on the theory by which these thermal reactions take place. It is true that all those reactions take place only at a very high temperature, and they are very complex. According to the schemes, when a hydrocarbon is subjected to heat at higher temperatures, multiple paths are available for the molecule to crack into various smaller molecules. The concentrations of various species: paraffins, olefins and aromatics are dependent on the circumstances under which the experiments were carried out. Eventually, at a very high temperature a molecule can decompose into carbon and hydrogen, forming carbon as a product. The decomposition of molecules is associated with higher activation energies depending on the bond types, such as C-C, C-H or H-H (50-80 kcal/mol).

Experimental evidence does support occurrence of various chemical species through free radical reaction mechanisms. Laidler and Polanyi [4, 8] do provide satisfactory qualitative analysis of the process. Table 2.3.1 illustrates an example of the kinetic data at various steps of the pyrolysis of a simple case of propane at 660 °C. It is to be noted here that as the carbon chain length increases, the C-C bond energy decreases. According to their data, the rate of decomposition of a free radical is much lower than the rate of stabilization of the free radical through extraction of hydrogen from another molecule. The kinetic data in Table 2.3.1 also show that the energy required to further break a radical (32-38 kcal/mol) into another radical and hydrocarbon is less than half of the energy required to break the parent hydrocarbon itself (82 kcal/mol). On the other hand, the energy required for a free radical to stabilize through extraction of hydrogen from its parent molecule (as the unreacted molecule is present in excess) is less than 10 kcal/mol. There is almost negligible energy transfer involved when two free radicals recombine together to terminate a radicalic reaction with a very high rate of reaction.

The propagation reaction mechanisms and their corresponding kinetic values also illustrate the expected product distribution in a commercial thermal processing of hydrocarbons. The rate of reaction is proportional to the formation of energized molecules and is therefore related to the concentrations of the final products in a first-order reaction.

Table 2.3.1. Kinetic data for the pyrolysis of propane at 660 °C (source Laidler)

Reaction mechanism			k (s^{-1})	E (kcal/mol)
Initiation reaction				
C_3H_8	⟶	CH_3* + C_2H_5*	1.6 x 10^{-9}	82.0
Propagation reactions				
C_2H_5* + C_3H_8	⟶	C_2H_6 + C_3H_7*	3.2 x 10^9	10.0
CH_3* + C_3H_8	⟶	CH_4 + C_3H_7*	8.9 x 10^9	8.2
C_3H_7*	⟶	CH_3* + C_2H_4	1.9 x 10^6	32.0
C_3H_7*	⟶	H* + C_3H_6	7.2 x 10^4	38.0
Termination reactions				
2 CH_3*	⟶	C_3H_6	4.5 x 10^{13}	0
2 C_2H_5*	⟶	C_4H_{10}	2.5 x 10^{13}	0

During Van't Hoff's period (see the chart in the Preface), important reactions were explained either by unimolecular decomposition or the nature of the activated complex of a bimolecular reaction through collision theory. However, Hinshelwood[1] was first to give an explanation for a complex molecule's changes in kinetics values through a radicalic reaction mechanism. This is because he assumed that the energy possessed by a complex molecule is distributed among the various moieties or the length of the side chain present, and the total energy of the molecule is the sum of the distributed energies, as explained below. Also, he assumed that the energy that is ultimately involved in the process may not come from all the moieties present but from some of them that are favorable under these conditions, where the energy has become localized in a particular bond or bonds that are to be broken during the thermal reactions.

For example:

The normal paraffinic C-C bond energy is weaker than the corresponding secondary or tertiary state of the hydrocarbons:

primary C-C bond energy < secondary C- C bond energy < tertiary C-C bond energy

The C-C bond energy weakens further when the paraffinic molecule becomes a cyclic bond. The energy weakens with the decrease in carbon ring structure in cyclo-paraffins from C_6 to $C_{3,}$ as shown below:

	C_3-ring	<	C_4-ring	<	C_5-ring	<	C_6-ring
Δ-Bond energy (kcal / mol)	~ - 10		~ - 6		~ - 1		~ -0.5

(approximate negative values)

When the paraffinic bond becomes olefinic with multiple bonds, the bond energy increases with the increase in bond numbers from double to triple, as follows:

	C-C	< C = C	< C ≡ C
Bond energy (kcal/mol)	~60	~100	~120
	(approximate values)		

The high bond energies for the double or triple bonds indicate that they are difficult to crack in thermal processing, but are quite reactive in other reactions, such as hydrogenations, halogenations, or polymerization. Similarly, benzene has bond energy of 80 kcal/mol, in between C - C and C = C but is extremely difficult to break due to its resonance structure.

According to Nelson[10] (see Preface), usually thermal cracking reactions start around 380°C, though it is highly dependent on the reaction time. At a particular temperature, the decomposition of a molecule starts after a critical time. As the time increases, the conversion also increases almost linearly with the increase in primary cracking reactions; however, the yield of the primary desired product decreases after a certain time with the increase in secondary reactions. The secondary reactions, with the formation of some unwanted product such as coke, increase rapidly with the increase in temperature. Nelson further reports that when the reaction is carried out in the liquid phase at a lower temperature range (380- 425 °C / 700 - 800 °F), the rate of cracking reaction almost doubles with each 15 °C / 20 °F rise in temperature. But at higher temperatures (450 °C – 600 °C / 850 °F-1100 °F) in the vapor phase reactions, the rate of cracking reaction doubles for an increase in temperature of around 20 °C (or 35 °F). Nelson mathematically defined the relationship with the equation:

$$C_i = C_o 2^{[(Ti - To)/x]} \quad \text{........ (2.3.3)}$$

Where,

C_i = yield for a given time at temperature T_i
C_o = yield for a given time at temperature T_o
x = number of degrees required to double the yield

Thus, Nelson's study clearly shows the obvious pressure effect on the rate of reaction. As pressure increases for a given temperature, some of the vapor phase material will go into liquid phase, and thus the rate of cracking reaction will change depending on the amount of material in the liquid phase vs. vapor phase.

As a cautionary note to readers, the literature data on the kinetics of thermal cracking of hydrocarbons are all over the map. The main reason is that the study of such reactions is of specific interest to researchers, and the results are highly dependent on the way those reactions were conducted and the data were interpreted. According to the present author's own experience in this area, major differences were observed between reactions done in a batch reactor system versus in a continuous reactor system. A large majority of the research has been conducted in the batch system in a smaller scale, even in a mini or micro scale. The major effect is the scale of operation, and predominant of all is the concentration and recombination of free radicals in a closed system. Pressure has consequences on the kinetic data, and especially in a batch system, the actual temperature and pressure are not close enough to the real reported values. As the cracking reaction proceeds with time, the recombination of free radicals is believed to dominate. Thus, the reported first-order rate constants are combinations of first- and second-order reactions. In all cases, the batch reactor kinetic values are lower than that of continuous data under the identical reported conditions.

Thus, there is an ongoing debate on the application of smaller-scale batch experimental kinetic data on the industrial-scale operation, whether it is global rate of decomposition of hydrocarbons or the rate of formation of a specific product. Whatever the case may be, the biggest difficulties encountered are: (i) experimental method, (ii) method of calculation, (iii) interpretation of the results, and (iv) experience of the researchers.

2.4 Thermodynamics of hydrocarbons

Thermal stability of hydrocarbons is dependent on two major factors:

i. Intramolecular bond energy, and
ii. Energy of formation of the hydrocarbon from carbon and hydrogen

Pauling first reported that the bond energies vary depending on their positions, and the C-C bonds are weaker than the C-H bonds in the same molecules. Energy required for the hemolytic scission of some typical C-H and C-C bonds in gas phase thermal processing of hydrocarbons is shown in Table 2.4.1 (source Pauling[3]).

Table 2.4.1. Energy of dissociation of C-H and C-C bonds in hydrocarbons (kcal/mol)

Hydrocarbon	C- H bond	C-C bond
CH_3 -**H**	101	
CH_3-CH_2 -**H**	98	
CH_3-CH_2-CH_2 -**H**	95	
CH_3-CH_3		83
CH_3-C_2H_5		81

As shown in Table 2.4.1, the energy required to break the C-H bond decreases with the increase in molecular weight of the hydrocarbons, whereas the variation of C-C bond energies is not significant enough from ethane to propane. From these data, one can propose that the initiation reaction step to form radicals in gas phase thermal cracking of paraffinic hydrocarbons may be dominated by the C-C bond, as compared to the C-H bond. However, for a large-sized molecule in the liquid phase, as in the case of heavy crude oil, the theory may not hold.

2.4.1 Calculation of thermodynamic data (source reference 11)

The basic data necessary for any thermodynamic calculations are:

$$\Delta H, \Delta S, \Delta G \text{ and } C_p$$

Where,

Standard heat of formation of hydrocarbons at 298K	ΔH_{298}
Absolute entropy at 298K	ΔS_{298}
Standard free energy of formation of hydrocarbons at 298K	ΔG_{298}
Heat of combustion	Cp

A thermal reaction follows the simple thermodynamic relationship of:

$$\Delta G^o = \Delta H^o - T\Delta S^o \qquad \text{.......... (2.4.1)}$$

- A reaction is thermodynamically possible if ΔG^o is negative
- The value of ΔH^o is negative for exothermic reactions and positive for endothermic reactions

The Cp value could be determined from the following relationship:

$$\frac{d\,\Delta Ho}{dT} = \Delta Cp$$

Where $$\Delta Cp = \Sigma\,(\Delta Cp)_{product} - \Sigma\,(\Delta Cp)_{reactant} \qquad \text{.... (2.4.2)}$$

The thermodynamic data of the thermal processing of hydrocarbons could be calculated by several methods. However, one of the best known and the simplest method could be the empirical way, through approximate values, where the calculations are done based on the thermodynamic contributions of each type of intramolecular linkage, assuming that the values are additive for the whole molecules, as follows:

$$\Delta H_{298} = \sum_{i=1}^{n} (\Sigma(\Delta H))i \qquad \text{......... (2.4.3)}$$

$$\Delta S_{298} = \sum_{i=1}^{n} (\Sigma(\Delta S))i \qquad \text{......... (2.4.4)}$$

$$\Delta G_{298} = \sum_{i=1}^{n} (\xi(\Delta G))i \qquad \text{......... (2.4.5)}$$

Where, ξ represents the contribution of a particular molecular linkage type or moiety.

Table 2.4.2 gives typical values of ΔH_{298}, ΔS_{298} and G_{298} of various linkages of hydrocarbons. By using the values of smaller linkages, one can predict the global contribution of a larger molecule, as described below.

Table 2.4.2. Energy values of various C-C and C-H linkages at 298K of various hydrocarbons (Source: Ref 11)

Bond type	$\xi(\Delta H)i$ kcal/mol	$\xi(\Delta S)i$ cal/mol/K	$\xi(\Delta G)i$ kcal /mol
Aliphatic hydrocarbons			
C—C (primary)	-10.11	27.42	-3.93
C—CH2(—C) (secondary)	-4.93	9.31	2.00
C—H	-1.80	-11.85	6.79
Olefinic hydrocarbons			
C=C (ene - 1)	27.1	7.87	25.24
C=C—C (ene-2)	18.5	17.0	23.6
C=C=C (diene)	51.2	21.5	55.3
Aromatic Hydrocarbons			
C—H	3.3	10.7	5.1
C—C	5.5	-4.6	7.3
Cyclic hydrocarbons correction factor	cal/mol	cal/mol/K	cal/mol
3- member	24.5	31.1	15.3
5- member	6.2	23.4	-0.7
6- member	150	15.4	-4.4

In order to do the calculation of global value of a molecule from the contribution of each moiety, one may go through the following steps:

i. Write down the whole molecular structure.

ii. Calculate the number of various linkage groups in the molecule involved.

iii. One may need to use correction factors for some particular types of groups, like cyclization reaction or any specific isomerization.

iv. Then sum up the values.

Laidler[8] had reported that a linear relationship exists between the free energy of activations from one homologous series of reactions to another. He showed that for a considerable number of reactions, a plot of ΔH° against $T\Delta S^{\circ}$ was found to be a straight line with a slope of unity. Nevertheless, it is true for a simple system, with simple substituents. For this reason, it is always easy to give a successful explanation for a relatively simple process. For a complicated system, a much more complex explanation is involved.

2.5 Types of cracking processes

All hydrocarbon molecules are generally stable below 350 °C at atmospheric pressure. However, the thermal stability of a hydrocarbon decreases with the increase in molecular weight of the hydrocarbon and the types of hydrocarbons. For example, paraffins have less stability than that of the corresponding olefins. Paraffins, with a straight chain saturated hydrocarbon, are less stable than the corresponding cyclic saturated hydrocarbons, naphthenes. On the other hand, the cyclic naphthenes are less stable than the corresponding aromatics. Stability is defined as the temperature at which the hydrocarbon starts cracking. A moderate paraffin usually will not crack significantly below 400 °C and an aromatic will not crack below 450 °C.

The cracking process can be divided into four major classes, as follows:

i. Thermal cracking
ii. Catalytic cracking
iii. Steam cracking
iv. Hydrocracking

Thermal cracking is usually carried out for smaller paraffinic hydrocarbons in order to convert them into olefins at a higher temperature and lower pressure. There is no catalyst involved, and the process is radicalic in nature, as discussed in detail in Chapter 3.

Catalytic cracking is mostly done for the longer chain hydrocarbons utilizing zeolite, a silica- alumina-based catalyst, at a moderate temperature and pressure. Unlike the thermal process, the catalytic process takes a very different path based on ion formation.

Steam cracking is very similar to the thermal cracking process carried out in the presence of steam with no catalyst at a high temperature and low pressure. As discussed in detail in Chapter 6, the biggest application of the steam cracker is in the petrochemical industries.

Hydrocracking is a process carried out to convert heavier aromatic hydrocarbons into smaller molecules. The process is done in the presence of high pressure hydrogen and hydrogenation catalyst. The aromatic hydrocarbons tend to form coke if treated either in a thermal process or in a catalytic process without hydrogen. In the refining industry, the process is used to saturate the bigger aromatic ring and break into smaller molecules to produce gasoline- or kerosene-range hydrocarbons and at the same time to remove unwanted metals, sulfur, and nitrogen from the product stream. A detailed description of hydrocracking is beyond the scope of this book; however, it is available in this author's previous book published by the same publisher (reference 2, Chapter 1).

2.5.1 Reactions in the presence of catalysts

The main subject of this book is thermal cracking of hydrocarbons and not catalytic conversion of hydrocarbons, which is a vast area by itself. Just for the sake of readers' having a background and comparison with thermal cracking, this section is restricted to a simple snapshot of catalytic cracking of paraffins; however, details are available elsewhere[12].

Cracking reactions, being endothermic, are thermodynamically favored at higher temperatures. Thus, the whole purpose of adding catalyst is to decrease the activation energy of the reactions and hence decrease the reaction temperature. The catalyst is a very complex area of research where the selectivity and activity of a catalyst can be controlled by the change of metals, various metal ratios, its shapes, its active sites, its pore size, and pore volumes, etc.

Acid catalysts are commonly used for cracking reactions, though several secondary reactions are also expected at the same time such as isomerization, alkylation, cyclization, polymerization, etc. Commercial catalysts are made of synthetic silica-alumina (10-20%) or silica-magnesia (30-35%). Those acid-treated materials are commercially known as zeolites or molecular sieves. They are used in a cylindrical or pellet shape, bigger-sized for fixed-bed reactors, and in a fine powder in fluidized-bed reactors. The biggest problem with using a catalyst in a thermal cracking process is the rate of deactivation due to the formation of coke on the surface of the catalyst. As the temperature of the reaction increases, the rate of deactivation also increases. In other words, the life of the catalyst becomes shorter and shorter. For a fixed-bed reactor, the process becomes expensive with shut-down of the system for replacing the catalyst. In this case, the reactor is cooled and the catalyst is removed, and then regeneration is carried out by air combustion, i.e., burning the coke deposit.

It has been reported [12] that the apparent activation energy of normal hydrocarbons decreases with the increase in chain length, and hence the catalytic cracking conversion increases with the increase in carbon number under identical reaction conditions.

Table 2.5.1 illustrates the values of the activation energies and the corresponding conversion of normal heptane, octane, and decane at 500 °C, with SiO_2-Al_2O_3-ZrO_2 catalyst. The activation energy decreases by about 4 kcal/mol from C_7 to C_8 hydrocarbons and decreases further by another 4 kcal/mol from C_8 to C_{10} hydrocarbons. On the other hand, the conversion almost doubles from C_7 to C_8 and then from C_8 to C_{10}.

Table 2.5.1. Activation energy and conversion values for normal paraffins in a catalytic process

Paraffins	Activation energy, kcal/mol	Wt% conversion
Heptane	29	8
Octane	25	18
Decane	21	42

The catalytic dehydrogenation of paraffinic hydrocarbons to olefins (discussed in Chapter 4) has the most important commercial application to date for the petrochemical industries (as discussed in Chapter 6). The dehydrogenation reaction is thermodynamically favored at a higher temperature and lower pressure. In a catalytic process, one can obtain almost the same conversion and yield of olefins, say, ethylene from ethane or butenes from n-butane, at a temperature about 50 °C to 80 °C lower than the corresponding thermal processes[12].

A major difference is observed between thermal and catalytic processes, when the paraffinic hydrocarbon chain length is greater than six carbon atoms. The higher hydrocarbons have the tendency for cyclization followed by aromatization in the presence of a catalyst, as shown below:

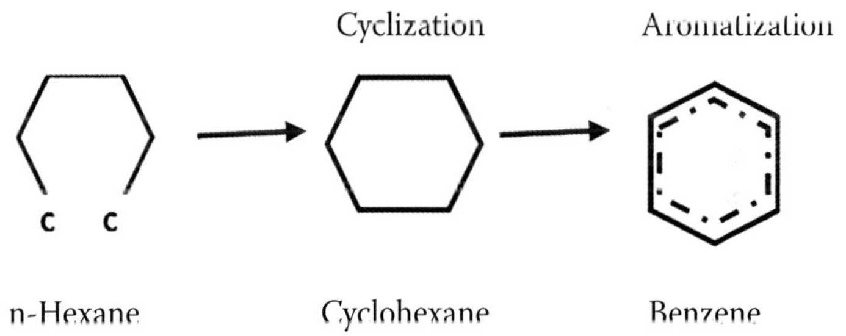

On the other hand, in a thermal process, the cracking reaction is predominant as compared to any other reaction. For that reason, dehydrogenation of higher paraffins (>C6) to their corresponding olefins is not favored by catalytic method in a commercial-scale operation.

References for Chapter 2

1. C. H. Hinshelwood published hundreds of papers in the Royal Soc of London in the 1950s and 1960s; a few examples are: p18 (1950), p486 (1950), p285 (1951), p20 (1952), p214 (1952), p339 (1953), etc.
2. H Eyring and K.J. Laidler, *Experimental methods in gas phase reactions*, MacMillan, London (1964).
3. L. Pauling, *The nature of the chemical bonds*, 2nd ed. Cornell University Press, New York (1955).
4. K.J. Laidler and J.C. Polanyi, *Progress in Reaction Kinetics*, vol 3, Pergamon Press, Oxford, (1965).
5. O.F. Rice & K.K. Rice, *The aliphatic free radicals*, Johns Hopkins Press, Baltimore (1975).
6. O. F. Rice & E. W. Steacie, *Atomic and free radical reactions*, 2nd ed., John Hopkins Press, Baltimore (1955).
7. A.F. Dickenson & J. Kerr, *Progress in Reaction Kinetics*, Pergamon Press, London, (1961).
8. K.J. Laidler, *Chemical Kinetics*, McGraw Hill (1965).
9. S. Arrhenius, Z. *Physik*, vol 4, 226 (1889).
10. W. L. Nelson, *Petroleum Refining Engineering*, McGraw-Hill Book Co., New York (1936).
11. Editors: Daniel R Stull, Edgar F Westrum, and Gerard C Sinke, *The chemical thermodynamics of organic compounds*, J. Wiley, New York (1969).
12. J. E. Germain, *Catalytic conversion of hydrocarbons*, Academic Press, New York and London (1969).

3

Gas-phase reactions of hydrocarbons / Pyrolysis

3.1 The chemistry of pyrolysis

Thermal processing of hydrocarbons in the gas phase at a very high temperature and lower pressure under inert atmosphere is called "pyrolysis." The process is widely used in the petrochemical industries to produce olefinic feedstocks for the polymer industry.

The majority of the gas-phase studies have been devoted to the thermal processing of lower paraffins only, probably due to the complexities of the aromatic hydrocarbons. Kinetic studies of thermal reactions are quite difficult due to the number of successive and competitive reactions involved. The general cracking reaction involves the rupture of the C-C bond, which is thermodynamically favored at a higher temperature and could be represented as follows:

(i) Cracking of paraffins:

$$C_xH_{2x+2} \longrightarrow C_yH_{2y} + C_zH_{2z+2}, \text{ where } x = y + z \qquad \text{..... (3.1.1)}$$

(ii) Cracking of olefins:

$$C_yH_{2y} \longrightarrow C_pH_{2p} + C_qH_{2q}, \text{ where } y = p + q \qquad \text{...... (3.1.2)}$$

(iii) Cracking of aromatics:

$$Ar - C_nH_{2n+1} \longrightarrow Ar\text{-}H + C_nH_{2n} \qquad \text{.................. (3.1.3)}$$

The major products are always the olefins of lower molecular weight than the parent hydrocarbons. The process continues with further breaking up of the product paraffins and olefins, followed by dehydrogenation and isomerization reactions. At higher temperatures, polymerization of the olefins takes place with the formation of molecules larger than the parent hydrocarbons. In the case of aromatics, in particular, the aromatic free radicals suffer extensive polymerization and subsequently form coke through a carbonization reaction.

In earlier days, the molecular reaction mechanism was established before the free radical concept was introduced. Nowadays, the vast majority of the literature data are based on the radicalic reaction mechanism. The chemical pathway of cracking reactions and their product distribution could be better explained by the radicalic models proposed by Steacie, Hinshelwood, and other coworkers.

As described earlier, the free radical reaction proceeds randomly in chain as radicals start forming. Because the radicals have a shorter life, new radicals are always forming; thus, there is no accumulation of radicals and increase in their concentrations at a particular reaction time. However, after the cracking reaction, there is a tendency to form free radicals of the type that has the most stability or longest life.

3.1.1 Disproportionation reactions

The most probable reactions in gas-phase thermal processing is the interaction of a free radical with other molecules or radicals. It is possible for two radicals to undergo disproportionation reactions, as given by the following example:

$$C_3H_7^* + C_3H_7^* \longrightarrow C_3H_6 \text{ (olefin)} + C_3H_8 \text{ (paraffin)}$$

Along with recombination reactions:

$$C_3H_7^* + C_3H_7^* \longrightarrow C_3H_7 - C_3H_7 \text{ (paraffin)}$$

The disproportionation reactions actually could also take place after the formation of an excited hexane molecule, which then breaks down into propane and propylene:

$$C_6H_{14} \text{ (excited)} \longrightarrow C_3H_6 \text{ (olefin)} + C_3H_8 \text{ (paraffin)}$$

As it is difficult to measure the intermediate radicals, it is not certain which path is dominant.

3.1.2 Addition to double bonds

Another possible reaction that a radical can undergo is to add to an olefin. As Frey[6] reported, in this type of reaction the product itself is a radical while studying radicalic reaction with ethylene:

$$H^* + C_2H_4 \longrightarrow C_2H_5^*$$

Frey also reported that a foreign molecule is required in this reaction since excess energy must be removed for the stabilization of the free radicals. He also showed that there was formation of a cyclic compound by reacting methylene radicals with propylene, as follows:

The product is methyl-cyclopropane

3.2 Method of kinetic calculations

This subchapter describes a simplified method for the determination of "global order of reaction" during pyrolysis of the lower hydrocarbons. The basic principle is based on the equation proposed by Froment[1]:

$$x = f(\tau) \qquad \text{............ (3.2.1)}$$

Where,

'x' is the conversion of the feed hydrocarbons

and 'τ' is a function of the reaction time

If the reaction is carried out in a continuous tubular reactor and the reaction is homogeneous of the type:

$$F \longrightarrow P_i$$

Where,

F = feed and

P = product at any instant 'i'

the global kinetic rate constant 'k' of the order 'n' could be represented by the following equation:

$$k = \frac{1}{Co^{n-1}} * \frac{1}{\tau}\int_0^{xi} [(1 + \alpha \cdot xi + \delta) / (1 - xi)]^n \, dx_i \qquad \text{....... (3.2.2)}$$

Where,

Co = initial concentrations of hydrocarbon (mol / l)

δ = dilution factor (number of moles of hydrocarbon / mole of diluent)

α = expansion factor (number of moles produced / mole feed)

τ = reaction time (sec)

n = global order of reaction

x_i = global conversion factor of the feedstock at time 'τ'

Due to the difficulties in solving such an equation, the order of equation is determined by the best fit method, utilizing various orders such as 0.5, 1.0, 1.5, 2.0, etc. The results are usually not valid for large variations of conversions or large variations of reaction times. The best fit curve x = f (τ) is obtained for the lowest value of 'n' for which the 'k' value remains constant with time.

3.3 Pyrolysis of aromatic hydrocarbons

In earlier days, due to the lack of experimental data, there was considerable interest in the kinetics of the pyrolysis of aromatic hydrocarbons involving free radicals. The present author studied the kinetics of the pyrolysis of alkylbenzenes in detail, and some of the results of these investigations are discussed next[2a & 2b]. The basis of the work was to investigate the factors that influence these types of radicalic reactions. It provides a means of producing radicals with excess of energy so that the mechanism of their decomposition could be studied.

Experiments were carried out in a continuous tubular quartz reactor of 8 mm diameter between 600 to 700 °C, at atmospheric pressure, in an inert medium of nitrogen. Nitrogen was mixed with the aromatic hydrocarbons in 1:1 molar ratio inside the preheating zone. The gaseous and liquid products were collected separately and analyzed by gas-chromatography. Major gaseous products were C_1 to C_4 paraffinic and olefinic hydrocarbons, which includes butene-1, cis- and trans- butene-2, and butadienes. The major liquid products were toluene, xylenes, and styrene, along with relatively minor components of benzene, methyl styrene, diphenyl, bibenzyl, anthracene, phenyl butenes, phenyl butadienes, tetraline, and naphthalene.

Tables 3.1 shows the first-order rate constant values of the pyrolysis of alkyl aromatics calculated based on the method proposed by Hougen and Watson[3], and Table 3.2 represents data based on Froment's equation[4]. Kinetic calculations were done assuming that the reaction is homogeneous and of first order, taking into consideration the dilution factor and expansion factors, as given by Hougen and Watson.

Table 3.1. First-order rate constant values of the pyrolysis of alkyl aromatics

Aromatic Hydrocarbon	Temperature, °C	Conversion fraction, X_i	Rate constant k, (s^{-1})
Benzene	700	0.08	0.018
Toluene	700	0.10	0.028
Xylene	700	0.12	0.058
Ethyl Benzene	600	0.51	0.045
	650	0.48	0.411
	700	0.69	1.313
n-Butyl Benzene	600	0.49	0.262
	650	0.43	0.581
	700	0.63	2.865

The kinetic results in Table 3.1 are consistent with the radicalic mechanism[2b]. The first-order rate constant 'k_1' values are the highest for n-butylbenzene, whereas benzene has the lowest values. Benzene being the most difficult molecule to break due to its resonance structure, the only option benzene has to break the $C_{aromatic}$-H bond is to produce phenyl radical. The most probable further reaction is to recombine two phenyl radicals to form diphenyl, and that is the major product from the pyrolysis of benzene. The difficulty in initiating a cracking reaction in benzene is shown by the fact that less than 10 wt% maximum conversion of benzene was obtained at 700 °C.

The rate constant values of the alkyl-aromatic hydrocarbons increase with the increase in the length of the alkyl groups from toluene to ethyl benzene to n-butyl benzene. Toluene and xylenes have k_1 values only 2 to 3 times more than that of benzene, but when the aromatic compounds containing an alkyl group of beyond one carbon atom, such as in the case of ethyl (C2) and butyl (C4) benzenes, the rate constant values increased drastically by more than 200 times at 700 °C. The results could be clearly explained by the "free–radical" reaction mechanisms, as follows:

While the initiating reaction for benzene is the cleavage of the most difficult $C_{aromatic}$-H bond (bond energy 465-470 kJ/mol), for toluene and for xylene, it is the scission of the C-H bond of the methyl group (400- 410 kJ/mol) attached to the benzene ring structure. On the other hand, for ethylbenzene and beyond, the initiating reaction is the cleavage of the C- C bond, which has much lower bond energies (340-345 kJ/mol) than that of the C-H bonds adjacent to the aromatic nucleus. In the same study, it was observed that for ethyl benzene and beyond, for all alkyl aromatic hydrocarbons, the rate of reaction decreases with the increase in conversion at 700 °C. This clearly indicates that there is self-inhibition of the pyrolytic reactions, and it was postulated that olefins were most probably responsible for the inhibition in alkyl aromatics. This theory is in agreement with the results reported by Froment[4] and Frey[5, 6] that the styrenes are known to be a good inhibitor during the pyrolysis of paraffinic hydrocarbons.

The results in Table 3.2 were calculated based on Froment's best fit equation in order to determine the actual order of reactions instead of assuming first-order kinetics. The present interpretation has been done taking into consideration the cause of the product formation during thermal cracking is by the cleavage of multiple types of bonds and radical recombination reactions. But it has been impossible to determine at which degree these reactions are mainly taking place. It is assumed throughout this investigation that the reaction obeys the theory of first-order reaction kinetics. However, in the present case, attempts were made to determine the best fit order of reactions. It is clearly demonstrated in Table 3.2 that the order of reaction decreases with the increase in molecular weight of the aromatic hydrocarbons. For ethyl and n-butyl benzenes, where order of reaction data are available at two different temperatures, it has also been observed that the order decreases with the increase in temperature and reaches closer to unity at higher temperatures. Thus, it could be postulated that at lower temperatures, most probably recombination of free radicals is very active towards the formation of products.

Table 3.2. Global order of reaction for the pyrolysis of aromatic hydrocarbons

Aromatic hydrocarbon	Temperature, °C	Order of reaction, n	Rate constant, k $(mole/l)^{1-n} * s^{-1}$
Benzene	700	2.15	2.6×10^{-6}
Xylene	700	1.30	0.07
Ethyl Benzene	650	2.05	1.68×10^{-4}
	700	1.55	0.11
n-Butyl Benzene	600	1.55	0.07
	650	1.10	0.33

3.4 Pyrolysis of lower hydrocarbons (C_1 to C_4)

3.4.1 Methane

Methane being the first member of the hydrocarbon homologous series with one carbon atom, it has the highest C-H bond energy (104 kcal /mol or 435 kJ /mol) and thus the most stable hydrocarbon. Methane can be thermally decomposed only above 800 °C, and the thermal reaction is bimolecular with order of reaction $n > 1$. The major component after the initial methyl free radical formation is mostly acetylene, and the most acceptable theory for the steps involved is as follows:

$$CH_4 \longrightarrow CH_3^* + H^* \qquad {}^*\text{Free radical}$$

$$2CH_3^* \longrightarrow C_2H_6$$

$$CH_4 \longrightarrow C_2H_6 \longrightarrow C_2H_4 \longrightarrow C_2H_2 \longrightarrow 2C + H_2$$

$$CH4 \longrightarrow H_3C-CH_3 \longrightarrow H_2C=CH_2 \longrightarrow HC\equiv CH \longrightarrow C + H_2$$

When methane was pyrolyzed between 800-900 °C, evidence for the formation of coke deposit was observed on the internal surface area of the reactor walls. It was hypothesized that the coke formation could only take place through the intermediate acetylene formation that could decompose rapidly into carbon and hydrogen at above 800 °C. Nevertheless, when the methane pyrolysis was carried out above 1000 °C, the major product was acetylene and not ethylene, due to the fact that acetylene is more thermally stable than ethylene under those conditions. Thus, methane is not a suitable feedstock for the production of ethylene.

3.4.2 Ethane

Among all the gaseous hydrocarbons, considerable amounts of research resources have been devoted to the thermal processing of ethane. As discussed in Chapters 4 and 5, the main reason is the increasing demand of ethylene in the petrochemical industries for the production of polyethylene plastics.

Various sources reported that the yield of ethylene decreases in the presence of hydrogen. That confirms the fact that the formation of ethylene is due to the direct dehydrogenation of ethane through ethyl radical, instead of via intermediate formation of methyl radicals, as proposed by several other researchers. The theory of dehydrogenation has merit due to the fact that the formation of ethylene follows the first-order reaction. The various product distributions from the pyrolysis of ethane could be explained through the following mechanisms:

$$C_2H_6 \longrightarrow 2CH_3^*$$

$$CH_3^* + C_2H_6 \longrightarrow CH_4 + C_2H_5^*$$

$$C_2H_5^* \longrightarrow C_2H_4 + H^*$$

$$C_2H_6 + H^* \longrightarrow C_2H_5^* + H_2$$

$$2\,C_2H_5^* \longrightarrow C_2H_4 + C_2H_6$$

$$2\,C_2H_5^* \longrightarrow C_4H_{10}$$

$$2\,C_2H_5^* \longrightarrow CH_4 + C_3H_6$$

Table 3.4.1 shows a typical product distribution from the pyrolysis of ethane at the highest conversion level of 60 to 65%. The primary products were ethylene and hydrogen (30% to 35%), while the secondary product was methane (<5%). Relatively smaller quantities of less than 1% C_3 / C_4 hydrocarbons were formed.

Production of methane confirms the fact that there was cleavage of C-C bonds in ethane that forms intermediate methyl radicals. However, the dehydrogenation reaction dominates over the cracking reaction.

It is important to mention here that the above experiments were done in the laboratory by the author in a small-scale fast-pyrolysis reactor with single pass of the feedstock. However, in commercial operations, hydrocarbon feedstock is blended with steam and usually feedstock is recycled to increase the conversion. One can obtain maximum conversion of ethane in a lab scale between 60 to 65%, but in a commercial operation one could increase the conversion to about 80%.

Table 3.4.1. Product distribution from the pyrolysis of ethane

Components	Mol%
H2	30 - 32%
Methane	4-5%
Ethylene	32- 30%
Ethane	33 - 31%
C3 hydrocarbons	1 -1.5%
C4 + hydrocarbons	0.5 -1.0%

Figure 3.4.1 shows how the ethane conversion varies with the variation of temperatures and times. At 700 °C, the conversion is below 50% at the highest residence time of 3 seconds, whereas with the increase in temperature to 900 °C, almost the same conversion was obtained within 0.1 second. Similarly, about 60% conversion was obtained at 800 °C in one second and at 900 °C in less than half a second.

However, increasing the severity in order to obtain higher conversion is not an economical solution in a commercial operation. At higher severity, the selectivity towards ethylene decreases as it dehydrogenates to acetylene and eventually converts into carbon that deposits on the surface of the furnace wall as coke.

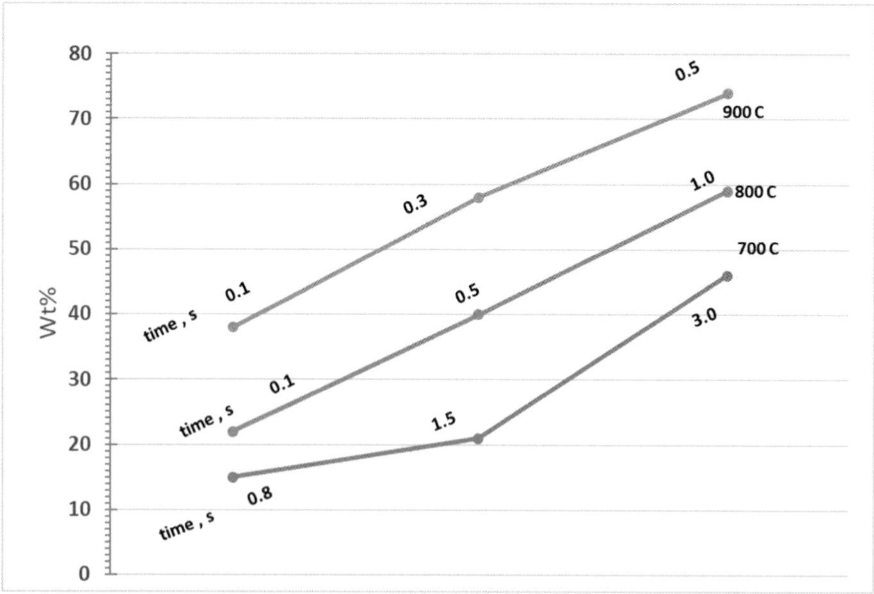

Fig. 3.4.1. Variation of ethane conversion with time and temperature

3.4.3 Propane

Pyrolysis of propane is usually carried out under atmospheric pressure at a lower temperature (below 700 °C) than that of ethane. The major products obtained from the decomposition of propane could be explained by the following reaction mechanisms[7]:

$$C_3H_8 \longrightarrow C_2H_4 + CH_4 \quad \text{............ (3.4)}$$

$$C_3H_8 \longrightarrow C_3H_6 + H_2 \quad \text{............ (3.5)}$$

$$2\,C_3H_8 \longrightarrow C_2H_6 + C_4H_{10} \quad \text{............ (3.6)}$$

The typical product distribution from the pyrolysis of propane at the highest conversion level (80-85%) is shown in Table 3.4.2. Free radical reaction mechanisms for the formation of the products could be explained as follows:

$$C_3H_8 \longrightarrow C_2H_5^* + CH_3^*$$

$$C_3H_8 + CH_3^* \longrightarrow CH_4 + C_3H_7^*$$

$$C_2H_5^* \longrightarrow C_2H_4 + H^*$$

$$C_3H_7^* \longrightarrow C_3H_6 + H^*$$

$$C_2H_5^* + H^* \longrightarrow C_2H_6$$

$$C_3H_7^* + CH_3^* \longrightarrow C_4H_{10}$$

Table 3.4.2. Product distribution from the pyrolysis of propane at 700 °C

Components	Mol%
H2	10-12%
Methane	35-36%
Ethylene	25-30%
Ethane	8-5%
Propylene	10-11%
Propane	8 - 5%
C4 + hydrocarbons	5-3%

The above reaction mechanisms explain the reason why methane and ethylene are the major products from the pyrolysis of propane. It shows that the primary C-C bond cleavage in propane is the dominating reaction with the formation of methyl and ethyl radicals, which are converted into methane and ethylene. The dehydrogenation reaction of propyl radical for the production of propylene is overall a bimolecular reaction, thus resulting in lower propylene yield. As shown in the reaction mechanisms, the formation of ethane and C_4 hydrocarbons are bimolecular reactions, thus yields of those products are lower too.

Thus, ethylene is a major product from propane as a feedstock at above 700 °C. Every mole of propane produces 0.46 mole of ethylene. At higher temperatures, the product propylene undergoes further decomposition with the additional production of ethylene. However, there is a maximum limit of temperature beyond which the process is not economical.

3.4.4 Butane

Steacie and Rice[7] reported that the thermal decomposition of butane takes place based on three major steps, as follows:

$$C_4H_{10} \longrightarrow C_3H_6 + CH_4$$

$$C_4H_{10} \longrightarrow C_2H_4 + C_2H_6$$

$$C_4H_{10} \longrightarrow C_4H_8 + H_2$$

(i) in the first step, major products are methane and propylene.
(ii) in the second step, the products are ethane and ethylene.
(iii) and in the third step, the products are butenes.

When the thermal processing of butane is carried out at a higher severity, almost 50% of the reaction takes place in the first step and another 35% in the second step.

The above products distribution could be explained through the free radical reaction mechanisms, as follows:

$$C_4H_{10} \longrightarrow 2C_2H_5^*$$

$$C_4H_{10} \longrightarrow C_3H_7^* + CH_3^*$$

$$C_2H_5^* + C_4H_{10} \longrightarrow C_4H_9^* + C_2H_6$$

$$C_4H_9^* \longrightarrow C_3H_6 + CH_3^*$$

$$C_4H_9^{*} \longrightarrow C_4H_8 + H^{*}$$

$$C_2H_5^{*} \longrightarrow C_2H_4 + H^{*}$$

$$C_3H_7^{*} \longrightarrow C_3H_6 + H^{*}$$

$$CH_3^{*} + H^{*} \longrightarrow CH_4$$

In butane pyrolysis, the production of ethylene increases almost linearly with the increase in butane conversion. It is possible to obtain maximum conversion of butane between 85% to 90%, as compared to 80-85% for propane and 60-65% for ethane.

Pyrolysis of C_4 hydrocarbon is carried out even at a lower temperature than the propane at around 600 °C.

The yields distribution in Table 3.4.3 shows that during the pyrolysis of butane, the dehydrogenation reaction is not the principal reaction, and cracking reactions are dominating. However, four to five times higher yields of ethylene as compared to ethane could not be explained based on the above reactions only. Thus, it is postulated that the following additional 4th reaction takes place during pyrolysis of n-butane:

$$n\text{-}C_4H_{10} \longrightarrow 2C_2H_4 + H_2$$

It is important to mention here that during pyrolysis of pure iso-butane, the production of ethylene is quite low, indicating the fact that the 4th type of reaction does not take place from the decomposition of iso-butane.

Table 3.4.3. Product distribution from the pyrolysis of butane at the highest conversion of 90+%

Components	Mol%
H2	11-12%
Methane	30-32%
Ethylene	20-24%
Ethane	4-5%
Propylene	14-12%
Propane	4-2%
C4 olefins	3-2%
C4 - paraffins	5-4%
C5 + hydrocarbons	2-3%

3.5 Pyrolysis of hydrocarbons in the presence of steam

In a commercial operation, olefin production is carried out through thermal processing of hydrocarbons in the presence of steam. The dilution factor, also known as steam to hydrocarbon ratio, is an important parameter for the process. The conversion of feed and hence the yields of olefins increase with the amount of steam introduced. The amount of heat input also increases with the increase of steam. Addition of steam also decreases the amount of coke formation or fouling inside the furnace wall. However, in the case of the pyrolysis of gaseous hydrocarbons, maximum 50% volume of steam with respect to the feed is effective; beyond that, there is no economic benefit of utilizing steam.

On the other hand, in the case of thermal processing of liquid hydrocarbons, a steam to feed mass ratio of 1.5 to 2.0 was found to be the most effective for the increase in conversion and yields of the desired product.

A simplified block diagram of a steam cracker process is shown in Figure 3.5; however, additional details are given in Chapter 6. The basic principal behind the steam cracker is that the super-heated steam is added to preheat the feedstock inside the reactor packed with fluidized inert material that acts as heat transfer media. Usually, the super-heated steam needs to be heated beyond one thousand degrees centigrade in order for the process to be effective. The energy demand in this process is very high, and the cost of operation increases with the increase in steam to feed ratio. Thus, one goal of the process is to control the amount of steam fed with the conversion. The economics behind this process involve the delta conversion achieved or the increase in olefin production with the cost of the addition of steam.

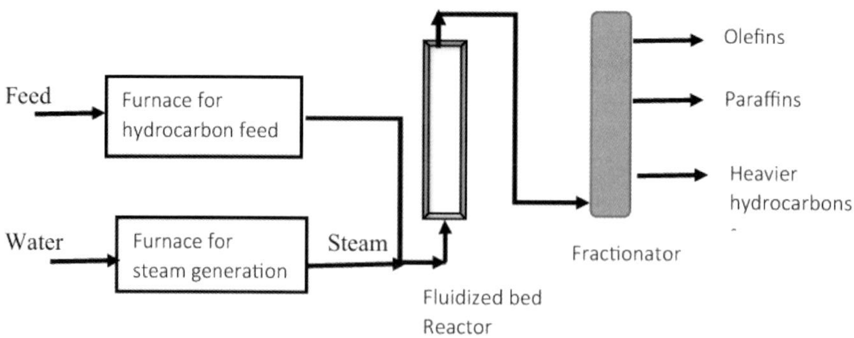

Fig. 3.5. Simplified block flow diagram of steam-cracker plant

3.6 Carbonization of aromatic hydrocarbons

The carbonization reaction during thermal processing of hydrocarbons is of considerable interest for the petroleum and petrochemical industries, as it has a practical application. Controlling the carbon deposit on the surface of the furnace tubing or catalyst surface or on the reactor wall during a commercial thermal cracking process enhances the desired product yield and increases the run length. Several efforts have been made to understand the mechanism of the carbonization reaction, which is basically the coke formation mechanism with the elimination of hydrogen, actually linked to the fouling.

As a general rule of thumb, the whole pathway for carbonization from paraffinic hydrocarbon as it happens in a petrochemical industry is shown in Figure 3.6.

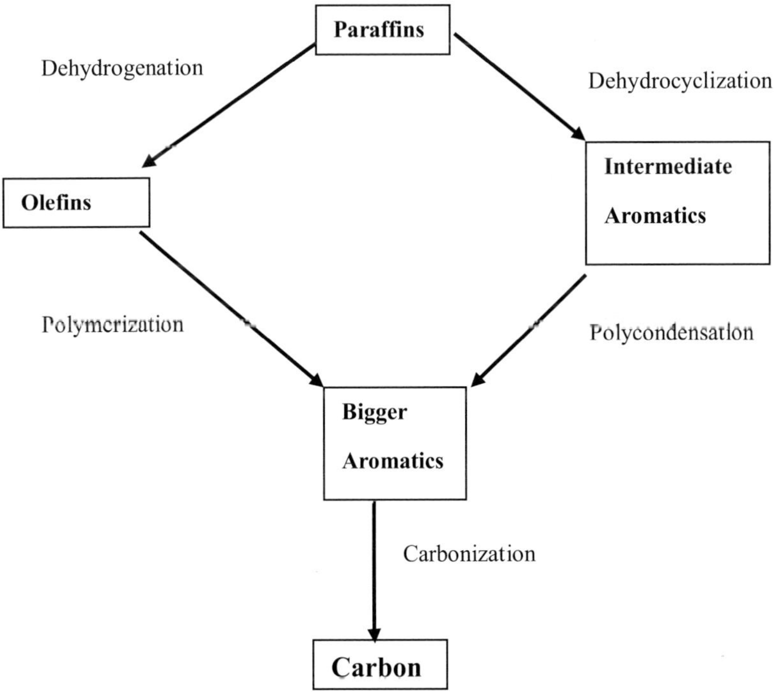

Fig. 3.6. A schematic diagram of the carbonization pathway

The petrochemical industries encounter serious fouling issues due to carbonization (see Chapter 6) during the dehydrogenation process of paraffins to olefins. As the severity of the reaction increases, the olefins formed are prone to polymerization and consequently to carbonization. Eventually, these reactions lead to the formation of a large aromatic structure that deposits as thin film on the catalyst surface and carbonizes itself, deactivating the active sites of the catalyst. In particular, the residence time at a given temperature plays a critical role in the carbonization reaction.

The carbonization reaction is promoted by the secondary cracking of olefins and random recombination of the free radicals thereby generated. It has been observed that the lower the atomic H/C ratio of the feedstock and the higher the temperature and residence time of the thermal cracking operation, the higher the rate of carbonization. The basic theory is that the carbon formation takes place with the total loss of hydrogen of the polynuclear aromatic hydrocarbons, the bigger aromatics, which are nonvolatile at the reaction conditions. Once the side chains are broken, the poly-aromatic free radicals selectively recombine towards the formation of coke. These reactions are favored at higher temperatures and pressures.

One of the first examples of studies of commercial application of the carbonization reaction of heavy hydrocarbons (not a model compound) was reported by Banerjee and Laidler[8]. They found that the first-order rate constant of the carbonization reaction decreases from left to right with the decrease in aromaticity or increase in H/C ratio of the hydrocarbon fractions, as follows:

$$k_1\ asphaltenes > k_1\ resins > k_1\ aromatics > k_1\ saturates$$

The kinetic studies of coking reactions of these complex mixtures of hydrocarbons were carried out by fractionation of heavy crude oil into their four fractions (SARA) mentioned above. Asphaltenes, having a poly-condensed aromatic structure, are the most aromatic in nature (lowest H/C ratio), followed by resins, which are similar to asphaltenes but of smaller molecular size. The aromatic fraction here is of smaller molecular size than that of asphaltenes and resins. Saturates having the highest H/C atomic ratio have the lowest rate constant values.

Styrene is an important petrochemical for the manufacture of polystyrene. The process is carried out in a fixed bed catalytic reactor, but they use a series of reactors in parallel, in order to switch the reactor when one catalyst bed is deactivated. Then, the reactor is cooled down, and the catalyst is regenerated by burning the carbon deposit on the catalyst surface and reused. The length of the catalyst life dictates the economics of the process.

Banerjee and Laidler[9a] studied the kinetics of thermal cracking of ethylbenzene for the formation of styrene. Results show that owing to the dehydrogenation of the side chain, styrene was found to be the primary product from ethylbenzene. However, the data in Table 3.6.1 illustrate that the rate of styrene formation decreases with the increase in conversion, mostly due to the formation of other byproducts. As the polymerization of styrene increases, it leads to the formation of larger hydrocarbons and eventually to carbon formation.

Table 3.6.1. Kinetic data for the formation of styrene from ethylbenzene and the rate of carbonization reaction

Temp. / °C	Conversion / wt%	*k Styrene / min⁻¹*	*k Carbonization / min⁻¹*
650	35.1	0.23	0.0022
650	48.3	0.12	0.0024
700	57.3	0.21	0.0038
700	69.7	0.17	0.0041

On the other hand, the results in Table 3.6.1 also indicate that the first-order rate constant value of the carbonization reaction almost doubled with the double in conversion of ethyl benzene. Decrease in the k_1 values of the styrene formation and increase in the rate of the carbonization reaction indicate that the polymerization of styrene proceeds at a faster rate with the increase in severity. Unfortunately, it is difficult to find the optimum conditions for the maximum yield as all the reactions are going in parallel. The economic benefit depends on the rate of fouling on the catalyst surface and also the fouling of heat exchangers.

3.7 Pyrolysis of coal

When coal is subjected to thermal treatment at elevated temperatures, the decomposition behavior depends on the type of coal (rank) and its properties. In the first step of the pyrolysis of coal, the volatile matter is discharged at about 300 °C before the actual decomposition of coal starts. There is a direct relationship between the rate of decomposition of coal and temperature. Earlier studies showed that the overall decomposition of coal follows the first-order kinetics with apparent activation energy varying from 200-330 kJ/mol[10]. Later on, the results were disputed on the grounds that the release of volatile matter in the earlier stage and the decomposition of carbon material in the coal are two different reactions. Release of the volatile matter is of low activation energy, and decomposition of the coal at higher temperatures is of higher activation energy.

Berkowitz[10a] has reported that the pyrolysis of coal is usually done between 500 °C to 600 °C, the primary product being the gaseous components with the additional yields of light oil and tar. Coal-derived liquid is highly aromatic in nature, and the aromaticity of the liquid product increases with the increase in temperature. The light oil mostly contains benzene, toluene, and xylenes. The gaseous products mostly consist of $C_1 - C_4$ hydrocarbons, and inorganic gases CO + CO_2, hydrogen, hydrogen sulfide, and ammonia.

Unlike petroleum coke, coal contains a large quantity of oxygen. Hence, coal tar contains quite high concentrations of oxygenated compounds such as phenols and nitrogenated compounds like pyridines.

Nevertheless, the pyrolysis of coal follows the same free radical reaction mechanism as that of the petroleum feedstock generated in the initial stage of the

3.9 Vapor-phase cracking vs. liquid-phase cracking

When a hydrocarbon undergoes thermal cracking reactions, the probability of the formation of a number of products increases with the increase in molecular weight of the feed hydrocarbon. As described in earlier chapters, the thermal reactions take place through the free radical reaction mechanisms. Thus, the probability of the formation of a particular hydrocarbon will depend on the thermal stability of the corresponding free radicals. For the formation of paraffinic hydrocarbons, the thermal stability of the alkyl free radicals are in the order of:

Tertiary > Secondary > Primary

Thus, the formation of a tertiary hydrocarbon is thermodynamically more favorable than that of its corresponding secondary hydrocarbon and followed by primary hydrocarbons, as shown below, for example, for the C_5 paraffins:

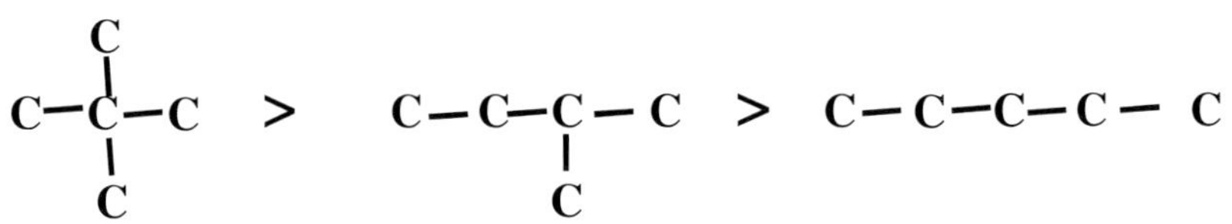

tertiary–Pentane secondary–Pentane normal–Pentane

In the case of aromatic hydrocarbons, the free radicals are stabilized through resonance, and it is known that benzyl radicals are more stable than phenyl radicals and more stable than any other longer-chain alkyl-benzenes; hence, the formation of toluene is more favorable than that of benzene during the thermal processing of alkyl-benzenes, as follows:

Ph – CH2* Ph*

Benzyl radical Phenyl radical

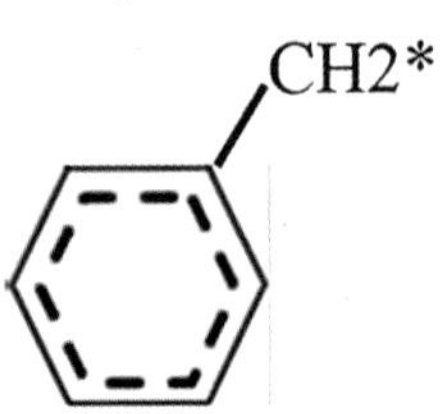

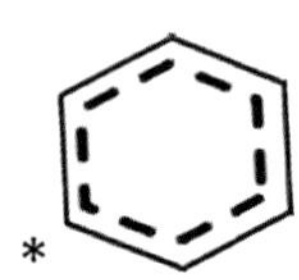

When any hydrocarbon feedstock undergoes thermal processing under identical conditions in the liquid–phase and in the vapor-phase, the reaction mechanisms of the formation of final products are different. It has been reported that the molecular addition and polymerization of olefins dominate in the liquid-phase with the formation of bigger molecules, whereas under the vapor-phase, cracking and dehydrogenation reactions dominate, with the formation of smaller molecules and olefins. Gray[14] and his coworkers showed that when n-hexadecane (C_{16}) was thermally cracked in the vapor-phase, about 5 times higher concentrations of C_1–C_4 hydrocarbons and olefins were formed as compared to the liquid-phase cracking under the identical reaction temperature. Much higher concentrations of > C6 hydrocarbons were produced in the liquid-phase, as compared to that of the vapor-phase cracking. Also, a significantly different addition reaction was observed under the liquid-phase reaction, with the formation of molecules bigger than the feed, whereas under the vapor-phase, no such products were observed, confirming the fact that the cracking reactions dominate in the vapor phase. Further, there was a significant decrease in the olefin concentrations in the liquid-phase, confirming the fact that the polymerization of olefins took place.

The same literature data also confirm that when the aromatic hydrocarbons undergo liquid-phase thermal reactions, the addition of olefins to the aromatic ring took place, with the formation of alkyl-aromatics due to the higher thermal stability of the benzyl radical, as discussed above.

3.10 Thermal reactions of sulfur compounds

Petroleum crude oil contains sulfur and nitrogen in various concentrations depending on the region from where it was recovered. As crude oil gets heavier, the amount of sulfur increases and at the same time the types of sulfur also vary, such as aliphatic sulfur, thiophenic sulfur, thiols, sulfides, disulfides, etc. The behavior of each sulfur type and their activities varies depending on their bond energies. It has been observed that the thiophenic S are mostly inactive under thermal cracking conditions, whereas sulfides and disulfides are quite reactive when heated to higher temperatures. Thus, it is reasonable to say that having sulfur, especially in heavy crude oil, favors overall cracking reactions, the reason being that C-S bond energy (310 kJ/mol) is less than that of C-C (aliphatic) bond energy (345 kJ/mol) or C-H bond energy (410 kJ/mol), and much less than that of C-C or C-H bonds in aromatic hydrocarbons. Even the S-S bond in the disulfide -C-S-S-C is weaker than that of C-S bonds and is expected to scission more easily and be first to promote the free radical initiation reaction, as follows:

(sulfide)	R-C-S-C	$\longrightarrow$	R-C-S* + C*
(disulfide)	R–S-S-C	$\longrightarrow$	R-S* + *S-C
(thiol)	R-S-H	$\longrightarrow$	R* + HS*

In spite of the benefit of cracking reactions mentioned above, having higher sulfur in a crude oil is not a desirable element. Eventually, the sulfur after cracking forms high sulfur lighter fuel (gasoline/diesel) and hydrogen sulfide. Then the gasoline and diesel have to undergo further severe hydrotreatment to meet the sulfur specifications. Hydrogen sulfide needs to be treated or scrubbed before the off gas is released into the atmosphere.

References for Chapter 3

1. G.F.Froment & Van De Steen, *J. Am. Inst. Chem. Eng.*, 23, 93 (1977).

2a. D. K. Banerjee Ph.D. Thesis, Institute of Petroleum and Gas, Romania (1980).

2b. D. K. Banerjee, V. Vintu and V. Matei, *Revue Roumaine de Chimie*, Vol 27, No 5, p 620-628 (1982).

3. O.A. Hougen and K. M. Watson, *Chemical Process Principles, Part III*, J. Willey, N.Y. (1947).

4. A.G. Buekens and G.F. Froment, *Ind. Eng. Chem. Proc. Des. Dev.* Vol 7, p 435 (1968).

5. H.M. Frey & R. Walsh, Thermal unimolecular reactions of hydrocarbons, *Chem. Review* (1968).

6. H.M. Frey, *Progress in reaction kinetics*, Vol 2 Pergamum Press, Oxford (1964).

7. O. F. Rice and E. W. Steacie, *Atomic and free radical reactions*, Johns Hopkins Press (1955).

8. D. K. Banerjee and K. J. Laidler, *Fuel* Vol 65, p 480-485 (1986).

9a. D. K. Banerjee and K. J. Laidler, *68th Canadian Chemical Engineering Conference*, Kingston, Ontario (1985).

9b. D. K. Banerjee and K. J. Laidler, Final report # 42105109, CANMET, Energy Research Lab, Ottawa, Ontario (1984).

10a. N. Berkowitz, *Fossil Hydrocarbons, Chemistry and Technology*, Academic Press, London (1997).

10b. N. Berkowitz, *An introduction to coal technology*, Academic Press, California, (1994).

11. R. C. Brown , *Why are we producing biofuels*, Publisher Brownia, USA (2012).

12. T. Carlson, T. Vispute, and G. Huber, *Green gasoline by catalytic fast pyrolysis of biomass*, Chem. Sus. Chem. 1, 397-400, (2008).

13. R. French, and S. Czernik, *Catalytic pyrolysis of biomass*, Fuel Processing Tech., 91, 25-32 (2010).

14. M.R. Gray, *Upgrading oil sands, bitumen and heavy oil*, University of Alberta Press, (2015).

4

Process design and development

4.1 Background

With the rapid increase in automation and high-tech plants, engineers should not believe that the plants can be designed simply with the help of a computer. It is important to have all the physical and chemical data of the process to be designed. For the best value from the design and development, the pertinent data necessary should be collected from a research laboratory with the existing pilot plant's experimental facilities part of the process. Preliminary data on mass and energy balances are desirable, with the optimum process conditions of temperature, pressure, and residence times. Data for the maximum feed conversion limit with the highest selectivity towards the desired product are necessary in order to obtain accurate design data.

Once the necessary data are available for the basic design and the straw-cost estimate with the approximate values, then a workable projection of a scale-up detailed design of the commercial value of the proposed project could be evaluated. Actual economics of a plant should be done only at the final stage as it is highly site-specific and based on several factors such as feed quality, product type and quality, reactor size and materials of construction, catalyst type and size, catalyst regeneration capabilities, plant location, etc.

The basic objective behind commercial production of any product is the economic aspect of it. In the petrochemical industry, the main challenge is the cost and availability of the feedstock, i.e., the paraffinic hydrocarbons. The next step is the value of the products, which may be olefins, aromatics, or any other intermediate sub-products. The key factor in the cost of production of olefins involves its fractionation and eventually the purity of the final desired product.

Considering the huge potential economic benefit to the petrochemical industry and at the same time stiff competition among various industries, it is worthwhile to discuss this topic here. As an example, it is important to know the various factors involved in the production of olefins for commercial application such as proper choice of feedstock and the corresponding technologies, marketing, and geographic and cultural aspects in various parts of the world. However, economists are constantly providing a prognosis for the bright future of these areas of commerce.

The main step is to understand the various reaction pathways involved in the process. For example, whether it is a vapor-phase reaction or liquid-phase reaction, what are the controlling factors of the process, do we have reliable experimental kinetics and thermodynamics data for the reactions involved, what are the yields and selectivities of the main product and of sub-products? In the thermal processing of heavier hydrocarbons, several side reactions and byproducts are generated that must be considered along with the market conditions.

The majority of publications are devoted towards commercial application of the olefins: ethylene, propylene, and butadiene. Rather less importance was shown towards isoprene, which is also drawing commercial importance recently. However, it is true that there is an increase in demand for ethylene for the production of polyethylene polymer and other poly-olefins. Another area of increasing interest is aromatics: benzene, toluene, xylenes isomers, and styrene. Styrene is a dominant feedstock for the production of polystyrene resins. The major challenge is the utilization of those primary products as feedstocks for the production of various other materials of commercial values such as polyethylene, polypropylene, elastomers, polymers, plastics, textile fiber, oxygenates, halogenates, fertilizers, etc.

The importance of these products will depend on the market demand, and it is expected to rise worldwide in billions of dollars in the future. According to a recent discussion at the Refining and Petrochemicals Meeting in the Middle East (Aug 2018), the global demand for petrochemicals is expected to increase more than 4%; on the other hand, the demand for transportation fuels is expected to grow only slightly more than 1% during the same period. That is because the growth will be diminished by the substitution of non-fossil fuel vehicles.

The success of development of any commercial plant is highly dependent on the constant supply of its feedstocks; hence, installation of a specific feedstock manufacturing plant within a transportable distance may be desirable. For example, a polymer manufacturing industry needs a nearby plant for the manufacture of lower olefins as their feedstocks; that in turn needs a steam generator for the steam cracker unit, a steam reformer for the production of hydrogen, a dehydrogenation plant to convert paraffins into olefins, and a fractionation plant for the separation and purification of the final product. In turn, one needs a nearby petroleum refinery for the availability of various petroleum fractions as feedstocks for the olefin manufacturing plant. At the same time, there are very few existing refineries in the world that are integrated with petrochemical plants in order to meet the above criteria. This provides access to the technology and feedstocks and also allows one to keep the operational costs under control through integration and to meet the market demand.

In addition to the above market, there are attractions for other chemical and petrochemical industries as well; for example, polycarbonates has a huge market in car glass, aircrafts, mobile phones, computer screens, etc. These are the top class petrochemical products.

It is true that no preliminary design of a process is complete without an economical evaluation. Early-stage economic calculation helps in the preparation of future decisions, but the final decision is not dependent on these early economics. In the course of the decision process, it is necessary to determine the most suitable economical process to obtain the desired product. As discussed in Chapter 6, there are several different olefin manufacturing processes available to market ethylene or propylene. In order to determine the most suitable process for that particular situation, various processes must be compared and evaluated.

One person cannot be expert in all the steps involved in a plant design. The purpose of this chapter is to point out the various steps involved, especially for the design of petrochemical and plastic industries. Nevertheless, for the success of a project, teamwork is necessary among various experts.

4.2 Markets for various olefins

4.2.1 Ethylene

Ethylene is the most demanding raw material in the olefin series for the plastic industry. There are various polymers such as polyethylene (PE), polyethylene terephthalate (PET), polyvinyl chloride (PVC), and polystyrene (PS), as well as other organic chemicals. Ethylene production and consumption are the highest in North America followed by Europe and then Japan.

4.2.2 Propylene

After ethylene, propylene is the most demanding feedstock for the petrochemical industries. Propylene may be produced either by dehydrogenation of propane or pyrolysis of iso-butane. Mostly propylene is used within the refinery as an alkylation agent for the production of alkyl benzene, in order to increase the octane value of gasoline. In the petrochemical industry, major productions are acrylonitrile, phenol acetone, iso-propanol, etc.

4.2.3 Butadiene

Commercially known butadiene is chemically known as 1, 3 butadiene. It is a colorless gas that is easily condensed as liquid for transportation and storage. Polybutadiene is mainly used for the production of synthetic rubber. The major source of butadiene is through dehydrogenation of butanes. In some refineries, it is recovered from the C_4 fractionation tower of the pyrolysis plant of lower hydrocarbons. The quality and quantity of butadiene recovered depends on the feedstock quality.

Butadiene can be produced from n-butane only above 700 °C. At 75% conversion level of n-butane, the mol % yields of butadiene, butene -1, and butene-2 are almost the same, at around 25 mol% each. However, if the temperature is increased further to 750 °C, butadiene formation is favored and almost doubled with a corresponding decrease in butene -1 and butene -2.

4.2.4 Isoprene

Commercially, isoprene is available as a byproduct of the ethylene industry. Though, it is produced from thermal cracking/steam cracking of naphtha, but the yield never exceeds more than 5% in the mixture of various olefins. Fractionation and separation of isoprene from the complex mixture of C_4 and C_5 olefins is quite difficult as the boiling points of all C_5 olefin isomers are very close to each other. Another option is solvent extraction of isoprene that can produce polymer-grade pure isoprene. More than 90% of isoprene is used for the production of poly-isoprene, which is a synthetic rubber. Natural rubber also consists of poly-isoprene of similar properties[1].

4.3 Feedstock Selection

Installation of an olefin manufacturing plant in a refinery is an economic issue; it depends not only on feedstock availability but also on geographic location. Especially for countries importing petroleum hydrocarbons, feedstock is of great concern. It is not only an economic decision but also relates to market demand, whether available lighter naphtha fraction should be used as a gasoline blend or as feedstock for a pyrolysis plant for the production of olefins.

As discussed in the previous chapter, methane is a highly stable molecule with the highest bond energy for the initiation of cracking reaction. Thus, methane is not a suitable feedstock for the production of olefins. Especially for the production of ethylene, it cannot be used, as ethylene is less stable than methane at the temperature of above 1000 °C required for the conversion of methane.

The quality of the feedstock is very important when utilizing liquid hydrocarbon as feedstock. It is to be taken under consideration whether the feed is paraffinic or naphthenic in nature. The severity of the operating conditions and the yields of ethylene depend on the types of feed properties[1]. For example,

(i) If the feed is highly paraffinic, the severity of the operation could be in the conversion level of 55 to 65%, and the corresponding yield of ethylene will vary from 25% to 30%.

(ii) If the liquid feedstock consists of higher concentrations of naphthenic hydrocarbons, then the thermal operation could be carried out at a higher conversion level of 60% to 65%; however, the corresponding maximum ethylene yield remains around 25%. At higher conversion, the ethylene production decreases.

Table 4.1 illustrates data for the production of various olefins from various feedstocks[1]. The yield of ethylene is closer to 80% from ethane itself. But the yield decreases sharply with the increase in molecular weight of the feedstock from propane to heavy naphtha. The production of propylene, though highest at about 20% with propane, remains almost constant at around 15% when feedstock molecular weight increases from C_4 and beyond.

The yields of higher olefins (C_4") remain very low, below 5%; however, the yield increases with the increase in molecular weight of the feedstock. Overall conversion of the feedstock decreases from ethane to butane and beyond.

Table 4.1. Average production of olefins from various feedstocks

Feed	Ethane	Propane	Butane	Light Naphtha	Heavy Naphtha
Conversion, wt%	85%	73%	70%	80%	65%
Products, wt%					
Ethylene	77%	42%	40%	31%	26%
Propylene	3%	21%	16%	16%	15%
Butadiene	1%	3%	3%	4%	4%
Butenes	1%	1%	7%	4%	5%
C5 + liquid	3%	6%	4%	24%	15%

Consequently, the economics of the thermal cracking/pyrolysis process in a refinery depends on selectivity of the desired product, yields of the byproducts, cost of separation of the byproducts, and also the demand for the byproducts. Thus, it is quite complex to determine the economic value, which also depends on the geographic location of the plant.

When higher molecular weight liquid feeds such as naphtha fraction are used in a pyrolysis plant, the process becomes more complex. The quality of the naphtha, whether it is paraffinic, naphthenic, or aromatic will depend on the source of the original crude oil. Thus, the quality of the C5+ liquid product yields will vary as it may contain valuable byproducts such as benzene, toluene, and xylenes (BTX), which will change the economics of the process.

As evidenced by the data presented in Table 4.1, more than 50% of the commercial ethylene production in the US is carried out using ethane as feedstock. By comparison, in Europe liquid hydrocarbons are used as major feedstock, where the pyrolysis plant is built next to a refinery, and part of the distillate is redirected to the pyrolysis plant. This may be part of the reason why gasoline and diesel are expensive in Europe.

4.4 Pyrolysis Plant

Building a pyrolysis plant is, of course, based on economics; however, investment in this plant is quite complex. Not only are availability and cost of the feedstock one of the most variable operating costs, there are many other interdependent factors. Flexibility of utilizing feedstocks and the ability to handle variable feeds is an important criterion. Technology also comes into play as a function of economics based on feedstock handling, conversion, yields or selectivity to the desired product, distribution of byproducts, separation of the byproducts, and finally, marketability of various sub-products.

The product distribution mainly depends on the furnace design. There is a limit on the design based on gaseous hydrocarbons versus liquid hydrocarbons or mixture of hydrocarbons as feedstocks. Conversion and yields directly relate to the furnace design, which in turn depends on the feedstock quality, as shown in Table 4.1.

4.4.1 Reactor Design

Since the early days of the pyrolysis plant or steam cracker, the most popular reactor design was the tubular reactor. The length of the tubular reactor may vary from 50 m to 100 m, and may be vertical or coiled. The tubular reactor has the flexibility of operating within a wide variation of temperatures, pressures, and feedstocks. Usually, the reactor is a serpentine coil, made of material resistant to higher temperatures of up to thousand degrees centigrade and inserted inside a furnace. In order to obtain maximum conversion and selectivity of the desired olefin production, the serpentine coil has to be maintained at a higher temperature, shortest residence time, and the partial pressure reduced to the minimum.

In a commercial plant, all those parameters are maintained as far as the economics of the process permits. Pushing to the highest temperature limit makes the choice of the reactor material difficult, as it has to be temperature and corrosion resistant material of high cost.

The main factor that limits the process variables is the deposit of coke inside the serpentine coil wall surface. In certain time intervals, depending on the rate of fouling, the process has to be shut down for decoking or cleaning of the reactor walls. The interval between the two decoking processes also dictates the economics.

4.4.2 Furnace Design

The heart of the thermal cracking process is the furnace that controls the temperature of the reactor and hence the reaction. The economics of the process are also controlled by the furnace operation. The cracking process takes place inside a long tubular reactor suspended inside the middle of the furnace (Figure 4.4.2). The furnace has two main sections: (i) radiant section and (ii) convective section.

A natural gas burner is located at the bottom of the furnace into the radiant section where the temperature is very high. Fuel gas is injected into the burner, and the flue gas leaving through the top provides heat through waste heat recovery to the convection section, where the inlets of the hydrocarbon feed and steam are located.

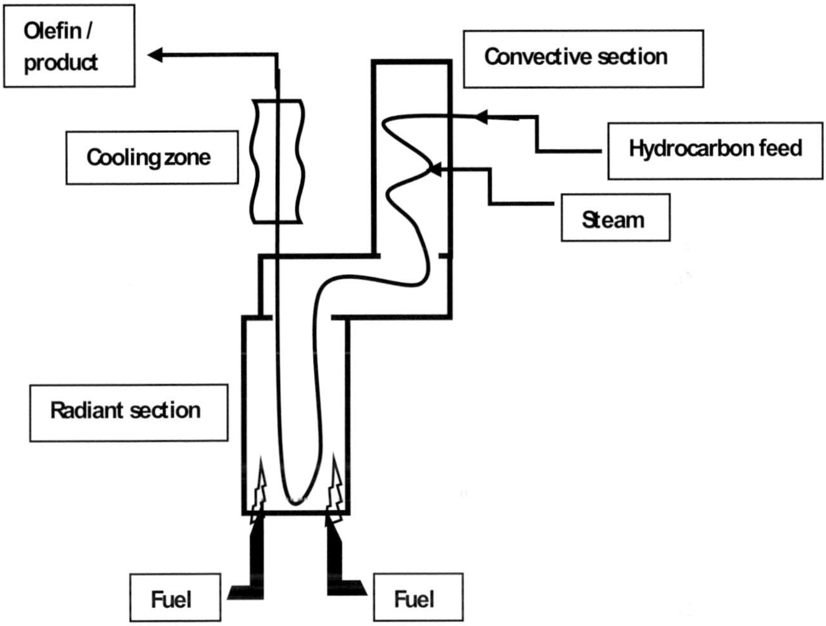

Fig. 4.4.2. Simplified diagram of a paraffin thermal cracker furnace design

Two important factors are considered for the furnace design: (i) rate of fouling inside the furnace tubing and (ii) run length. Maximum fouling usually occurs at the inlet of the furnace tubing and decreases along the length of the tubing. Also, higher fouling occurs at the hot spots. Thus, avoiding hot spots are an important consideration in the furnace design parameter. Run length is impacted by the number of days the process can be carried out between two de-coking operations.

The temperature of the furnace may vary from 650 °C to 900 °C. Length of the coil and the temperature of the furnace varies, depending on the feedstock. The maximum pressure is maintained just above atmospheric inside the reactor. The hydrocarbon feed is diluted with steam to increase conversion and prevent coke formation and backpressure buildup. Both hydrocarbon and steam are preheated in the convection section of the furnace (temp between 500 °C to 600 °C), before entering the radiant section where the temperature is between 700 °C to 900 °C, where the actual cracking takes place (source: Hydrocarbon Processing, March 2012).

Under these conditions, the saturated paraffinic hydrocarbons undergo dehydrogenation reactions with the production of unsaturated hydrocarbons like olefins, diolefins, and aromatics. As soon as the product stream leaves the furnace, it is quenched immediately to prevent unwanted secondary reactions like polymerization and coking. The cooling process is done in a specially designed heat exchanger.

It is important to avoid fouling, i.e., the coke formation inside the tubing, as far as possible for several reasons: coke is a bad conductor of heat; hence, it decreases the heat transfer from the furnace to inside the reactor wall. Because of that, it increases the fuel input to maintain the same reactor temperature, decreases the furnace run length, increases the pressure drop, decreases the internal diameter of the tubular reactor, hence causing decrease in the reactor volume, and subsequently changes the product yields. Eventually, the furnace has to be shut down. The de-coking of the reactor tubing may take more than 20 to 24 hours, and the longer the shutdown period, the more expensive the process becomes.

4.5 Process Variables

The product distribution from the pyrolysis of a typical hydrocarbon feedstock solely depends on the furnace design and its operating conditions. The most important process variables are:

i. Reaction temperature
ii. Residence time
iii. Partial pressure of the hydrocarbons

The process design consists of:

i. Mass flow rate
ii. Heat flux
iii. Internal diameter of the furnace tubing
iv. Pressure drop across the furnace length
v. Maximum temperature of the furnace wall
vi. Time interval between two decokings of the furnace

In general, all the process variables are strictly interdependent, and one variable influences the other factors. Thus, in order to obtain a desired product distribution or to achieve high selectivity of the targeted product from a given feedstock, change in only one variable is not enough to reach the target. Though the temperature in the serpentine coil plays the most important role, it is not constant. The temperature varies across the length of the furnace tubing; thus, the pyrolysis is not actually taking place at a certain temperature, but within a certain temperature profile across the length of the tubing.

The residence time of the reaction is interdependent of the temperature and in combination; it is the severity of the process that plays a crucial role in the conversion of paraffins into olefins. As discussed in earlier chapters, it is well established that the hydrocarbon conversion is a first-order reaction; hence, for a given severity the residence time is inversely proportional to the temperature and vice versa. Residence time is determined by analyzing the following factors:

i. Temperature profile
ii. Heat input
iii. Steam to hydrocarbon feed ratio
iv. Coefficient of expansion or increase in moles per mole feed
v. Partial and total pressure
vi. Length and internal diameter of the furnace coil

The heat input is dependent on the temperature profile, temperature of the furnace wall, and the steam to feed ratio. The pressure inside the furnace tubing is also not constant; it varies along the length of the tubing. Thus, the operating pressure is an average pressure across the furnace and is a function of feed rate, steam to feed ratio, and pressure drop across the tubing. This is calculated from the linear velocity and the dimension of the serpentine coil.

Once the furnace design is complete and constructed for a given pyrolysis plant, the olefin production can be controlled for a given feedstock only by operating variables such as temperature of the inlet and outlet of the serpentine coil, steam to hydrocarbon ratio and outlet pressure. Feed rate could be varied only within a limited range, and the overall effect on the severity and hence conversion is quite small.

There is no simple correlation between different variables in order to measure the efficiency of the thermal process. For a simple unimolecular system, the correlation between temperature and time could be expressed in terms of severity. Though there is no direct definition of severity, the "degree of severity" is expressed in terms of the "degree of conversion" of the hydrocarbon feedstock, which could be measured easily. On the other hand, in the case of a mixture of hydrocarbons, as in the case of pyrolysis of naphtha fraction, the actual conversion is not easy to determine.

4.6 Severity of thermal processing

The efficiency of a thermal process is compared for different feedstocks based on the degree of conversion of the feed or the selectivity of the desired product, which is defined as:

Conversion (weight %) = Grams produced per 100-gram feed

Selectivity = Number of moles produced per 100 moles converted

There is no severity scale to compare, as it is a combination of temperature and time, and may include pressure, too, for higher-pressure systems. However, the severity could be defined in terms of an arbitrary equation as dimensionless numbers as follows:

$$\text{Severity } \S = k * \tau \qquad \text{.............. } \mathbf{(4.6.1)}$$

Where,

'k' is the first order rate constant (s^{-1}) for the conversion of the feed and
'τ' is the reaction time in seconds taken for the conversion to achieve.

Here, 'k' is represented by the well-known Arrhenius equation:

$$\log (k) = \log (A) - E/ 2.3\, RT \qquad \text{....... } \mathbf{(4.6.2)}$$

Where,

'A' is the frequency factor (L mol^{-1} s^{-1})
E is the energy of activation (kcal/mol)
R is the gas constant (cal / mol. K) and
T is absolute temperature (°K).

One has to be careful in comparing temperature or residence time from one reactor to another, as there is no constant temperature along the length of the serpentine coil. Just for the sake of reference, one can use the temperature at the outlet of the furnace tubing. The rate of reaction also changes with the temperature; thus, the overall production of olefin is not constant throughout the length of the reactor. In a commercial plant, the overall conversion of the primary feedstock is increased through recirculation.

For a pure hydrocarbon such as ethane as feedstock, one can calculate with great accuracy the conversion of ethane and the selectivity of ethylene. However, for the pyrolysis of a mixture of hydrocarbons such as ethane and propane to ethylene, the calculation becomes more complex. The ethane remaining in the system is not only because of unreacted ethane but also consists of ethane produced from the propane. Thus, the calculation becomes more difficult when multiple hydrocarbons are present in the feedstock, such as in the case of naphtha cracking.

In those cases, the best option is to calculate simply the production of olefins based on the temperature, as shown in Figure 4.6.1. Results show the variation of yields of ethylene plus propylene combined from the pyrolysis of light naphtha fraction with the increase in temperature from 790 °C to 840 °C at a constant feed rate. Results show that the maximum production of olefins with average yield of 66% could be obtained between the temperature range of 820 °C to 835 °C. Thus, in a commercial plant, it is possible to obtain a constant production of olefins by maintaining the severity within a narrow range of operation (which includes other factors such as residence time too). However, with the increase in severity, the production decreases due to the polymerization of olefins and the carbonization reaction with the formation of coke.

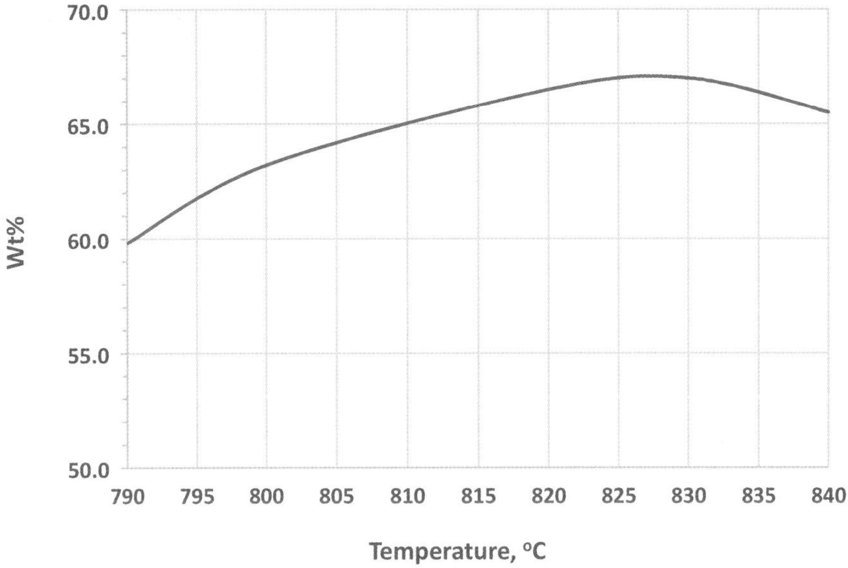

Fig. 4.6.1. Yields of C2+C3 olefins from the pyrolysis of light naphtha as a function of temperature

It is to be noted here that one can maintain the severity within a narrow range by varying the temperature and the residence time; however, the pressure plays an important role in the formation of olefins, too. Decrease in partial pressure of the hydrocarbon feedstock increases the production of olefins. In a commercial process for a given furnace, the material of the serpentine coil limits the maximum temperature, and the residence time is limited by the feed pump capacity. Running at a higher severity may not be economical, as the olefin production decreases and the coking reaction increases. In other words, the rate of fouling increases, which ultimately may shut down the production at a high economic cost.

As has been readily evidenced until this point, the pyrolysis of hydrocarbons is a complex process; however, in order to simplify the process variables, one can conclude the following:

i. The severity of operation of a steam cracker plant for the production of olefins is a function of temperature and time and they are interdependent, as shown in equations 4.6.1 and 4.6.2.
ii. The conversion of ethane into ethylene is directly related to the severity of operation; however, the production of olefin passes through a maximum, as at higher severity, olefin favors polymerization reactions and ultimately produces coke (Fig 4.6.1).

Though it is not easy to correlate severity in a commercial cracker operation, nevertheless, one can relate the severity of the operation, as explained below, using kinetic data (Figures 4.6.2 and 4.6.3).

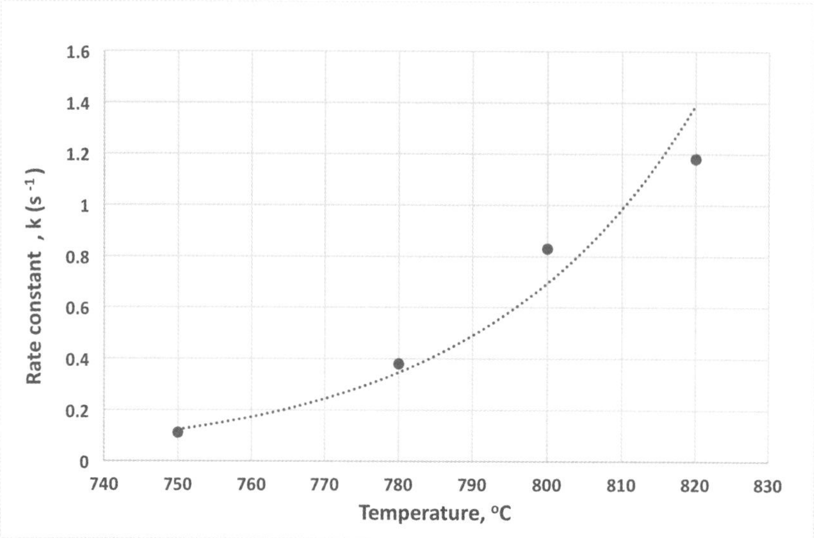

Fig. 4.6.2. Variation of first-order rate constant as a function of temperature of the pyrolysis of ethane

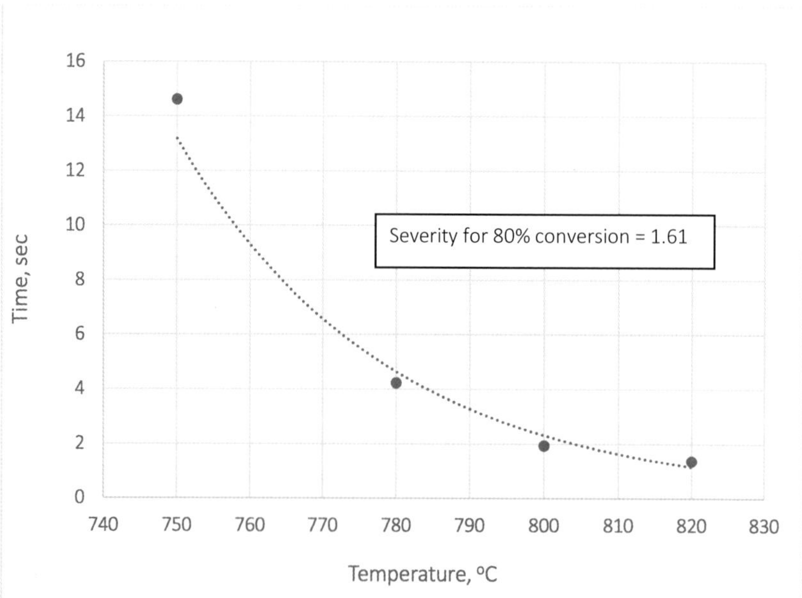

Fig. 4.6.3. Temperature-time relationship at the severity of 80% conversion of ethane

Figure 4.6.2 shows that the first-order rate constant values of ethane pyrolysis increase exponentially with the increase in cracker operating temperatures from 750 °C to 820 °C. Utilizing the kinetic severity equations (4.6.1), one can calculate at 80% conversion level the severity value is 1.61. In order to achieve the severity of 1.61, the temperature-residence time relationship is shown in Figure 4.6.3. The figure indicates that if the exit temperature of a serpentine coil furnace is maintained at 800 °C, the corresponding residence time required to achieve 80% conversion of ethane is about 2 seconds; if the temperature is dropped to 780 °C, the residence time is almost doubled to 4 seconds. In practice, in a commercial plant the above relationship could be used to determine the operating variables required for a desired conversion level, provided the kinetic data of the given feedstock is known. Though it is to be cautioned here that in a commercial plant, measuring exact residence time is difficult, and the temperature varies along the length of the serpentine coil furnace. Also during the process, the molecular weight of the components changes and hence the partial pressure along the length of the tubular reactor.

The formation of olefin is a dehydrogenation process; thus, the other option would be to relate the severity to the dehydrogenation of C_5 + material in a complex mixture of hydrocarbons such as naphtha as feedstock. That could be related to the H/C atomic ratio of the C_5 + material before and after the thermal processing. Thus, the H/C ratio of the C5+ material will decrease with the increase in severity. As the H/C ratio approaches unity, the severity is considered to be at the highest boundary limit and high enough to initiate a coking reaction at a significantly higher rate, which enhances fouling inside the furnace coil.

4.7 Selectivity

In general in the petroleum industry, the commercial production of a specific product is expressed in terms of yields, i.e., weight or volume percentage produced with respect to the feed. However, in the petrochemical industry, i.e., in the case of pyrolysis of hydrocarbons, yields may not be sufficient to express the success of a commercial operation. As we have seen above, especially in the production of olefins, the product yield decreases at higher severity. For example, simply in the case of butane pyrolysis, there are several products formed and the distribution of product slates changes with the increase in conversion of the feed or severity of operation. Thus, the term that is much more appropriate would be to express the production of a desired product in terms of selectivity, which could be expressed as:

$$\text{Selectivity} = \text{moles produced} \, / \, 100 \text{ moles converted}$$

In this case, one can determine at a certain conversion level the selectivity towards the formation of a desired product. In the case of butane pyrolysis, one can determine for a given conversion if the process is more favorable for the production of butene-1 or butene-2 (cis or trans) or 1,3 butadiene, etc. In the case of ethylene production, the selectivity towards the formation of ethylene from ethane decreases usually above 60% conversion level in favor of other products such as aromatization reactions or carbonization reaction with eventual coke formation.

4.8 Effect of partial pressure of hydrocarbons

For a given feedstock and at a particular severity of the cracker operation, the partial pressure of hydrocarbons plays an important role and is measured at the exit of the furnace tubing. However, it is not a simple calculation, as the partial pressure depends on several factors: (i) the total pressure, (ii) pressure at the exit of the furnace coil, (iii) loss of pressure along the length of the furnace tubing, and finally, on the (iv) steam to hydrocarbon ratio.

All the above variables are interdependent. The loss in pressure, however, depends on the length of the serpentine coil, internal diameter of the tubing, mass flow rate, and the density of the material inside the tubing.

Especially in the case of a mixture of hydrocarbons, such as naphtha cracking, the partial pressure plays an important role on the selectivity of the desired product, as it influences the rate of reaction and shifts the equilibrium of the reaction.

In the case of a unimolecular process, the production of olefins is inversely proportional to the partial pressure of the system; on the other hand, for a bimolecular reaction, the equilibrium of the reaction is relatively less affected by the partial pressure, depending on the type of reaction.

4.9 Steam to feed hydrocarbon ratio

In a commercial steam cracker operation, the steam to hydrocarbon ratio plays an important role, as discussed below:

i. Decrease in partial pressure of hydrocarbons
    a. indirectly increases the selectivity towards the production of olefins, a more valuable product for the plant.
    b. reduces the polymerization reactions of aromatic hydrocarbons, hence decreasing the formation of coke or fouling inside the furnace tubing.
    c. increases the time interval between two decoking processes, hence improving the process economics.
ii. Steam reacts with the metals (especially iron) in the tubing materials and hence decreases the rate of fouling inside the furnace tubing.

iii. Higher steam/feed ratio decreases the effective capacity of the cracker plant, i.e., decreases the amount of hydrocarbon fed per pass, hence decreasing the production of desired product.

Table 4.9 shows the ranges of steam to hydrocarbon ratios required in order to obtain optimum yield of the desired olefin product[1]. The data clearly indicate that the addition of a higher amount of steam is more effective for the higher molecular weight hydrocarbons. The results also show that in order to obtain a maximum production of olefins using heavier naphtha feedstock, the steam added should be more than 50% of the feed (quite high), while the steam cracker is running around 800 ° C. Steam to feed ratio is calculated based on the partial pressure required at the exit of the furnace coil. The steam/feed ratio could be increased further up to 0.8 or 0.9 in the event that the temperature is higher than 800 °C. In order to maintain the reactor temperature around 800 °C, the steam needs to be super-heated to a temperature between 1000 °C and 1200 °C. Another factor that controls the quantity of steam added is the residence time. In a serpentine coil reactor, the maximum recommended residence time for the production of ethylene is 1½ seconds.

Table 4.9. Optimum range of steam/feed ratios for various feedstocks in a commercial steam cracker

Feedstock, Hydrocarbons	Steam to feed ratio kg / kg feed
Ethane	0.30 – 0.35
Propane	0.35 – 0.40
Butane	0.40 – 0.50
Naphtha	0.50 – 0.70

However, in recent days, the direct thermal cracking process has progressively been replaced by the catalytic cracking process. The catalytic process takes place through intermediate ion formation, which could be cationic or anionic depending on the interaction of the paraffin with the type of catalyst. The catalyst could be Lewis acid-based for cationic reactions or alkaline metal-based for anionic reactions. The intermediate cationic reaction could be represented by the following example through heterolytic bond scission:

$$\mathrm{CH3 - CH2 - CH3} \xrightarrow{\text{catalyst}} \mathrm{CH3 - CH^{+} - CH3} \ + \ \mathrm{H^{-}}$$

$$\mathrm{H_2} + \ \mathrm{CH3 - CH = CH2} \longleftarrow$$

4.10 Separation of product gaseous hydrocarbons

Detailed discussion of the separation technologies is beyond the scope of this book. A brief overview is given below for the readers in order to understand the various steps involved in this quite complex system of separation process. There are several technologies available for the separation of lower paraffins from its olefins, such as cryogenic distillation, solid packed adsorption technique, membrane separation, solvent wash, etc.

There is always a constant drive in the research communities to develop new technologies that reduce energy utilization in the separation steps. There are some hybrid technologies available to reduce energy use and improve the degree of purity of the product. For the petrochemical industries, having the purity of their feedstock closer to 100% is extremely important.

In the petrochemical industry, the two most energy-intensive processes are steam cracker and paraffin/olefin separators. For the lower hydrocarbons (C_2 to C_4), the volatility and the boiling points of the paraffins are very close to the corresponding olefins, and the conventional purification through cryogenic distillation becomes very energy-intensive and difficult. However, the cryogenic technique can recover hydrocarbons at a very high purity.

The first step in the process of separating a mixture of lower gaseous hydrocarbons from the exit of the steam cracker plant or paraffin dehydrogenation plant (PDH) is to remove any free water and subsequently dry the remaining traces of water from the hydrocarbon stream. The second stage water drying process is usually done through a solid packed adsorbent column. The dehydrated mixture of gaseous stream is then fed to a compressor and chilled such that all hydrocarbons except methane are liquefied. After the separation of methane, all other hydrocarbons are separated by their carbon numbers in individual separators. Then olefin is separated from the corresponding paraffin in the splitter, and the paraffin is recycled back to the PDH unit.

Figure 4.10 illustrates a simplified schematic diagram of a cryogenic separation process for the step-by-step recovery of individual paraffins and olefins from a mixture of hydrocarbons at the exit of a steam cracker plant[2].

(i) Separator -1 represents a de-methanizer—where the mixture of gaseous steam cracker effluent is compressed to about 30-35 atmospheric pressure and enters at a temperature of -20 °C to -25 °C. Under these conditions, all hydrocarbons remain liquid except methane. Methane is separated from the top of the separator-1, and the bottom effluent is sent to the separator-2.

(i) Separator-2 represents a de-ethanizer, where the effluent from the separator-1 is slightly decompressed to about 20 atmosphere and temperature is maintained around -20 °C. Here, C_2 hydrocarbons are separated from the top and fed to the splitter-3, and the bottom stream is fed to the separator-4.

(iii) Splitter-3 separates ethylene from ethane. Pure ethylene is recovered from the top and stored, while the ethane from the bottom is recycled back to the PDH reactor.

(iv) Separator-4 is a de-propanizer, where C_3 hydrocarbons are separated from the rest of the heavier hydrocarbons by decompression at a higher temperature. C_3 is recovered from the top and fed to the splitter-5, while the C4+ hydrocarbons from the bottom are fed to the Separator-6.

(v) Splitter-5 is a propane-propylene separator, where propylene is collected from the top for further use, and propane is recovered from the bottom and recycled back to the PDH reactor.

(vi) Separator-6, is a de-butanizer, where C4 hydrocarbons are separated from the top and fed to the splitter-7, and the bottom stream of C5+ heavier hydrocarbons are collected and stored for future use.

(vii) Splitter-7 separates butenes from butane. The butenes mixture is sent to a different fractionator for the separation of butene-1 from butenes-2 and butadienes.

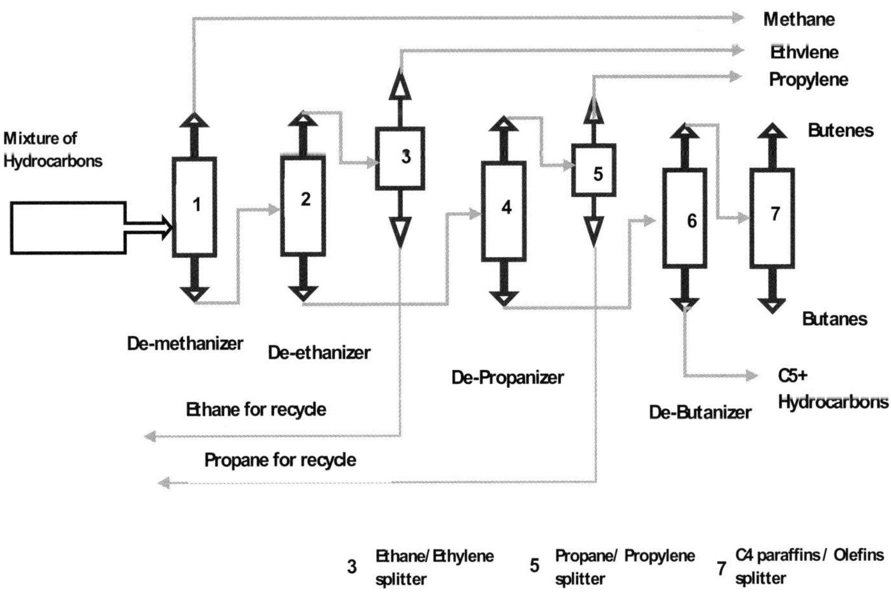

Fig. 4.10. A schematic diagram of the separation process of a mixture of gaseous hydrocarbons from a steam cracker plant

4.11 Polymerization of olefins

Strict environmental regulations in the last decade have affected the petrochemical industry, too. As a result, the plastic industry has been forced to develop environmentally friendly processes as well as products. The process of polymerization discussed in this sub-section is strictly related to the olefins, and the multiplication of C = C bonds for the production of polyolefins.

In general, a reaction of polymerization represents multiplication of monomer as follows:

$$\text{n (Monomer)} \longrightarrow \text{(Monomer)}_n \longrightarrow \text{Polymer}$$

In order to proceed with this reaction, the hydrocarbon should be unsaturated and should have at least one or more double bonds or triple bonds. As the degree of polymerization, i.e., the number 'n,' increases, the molecular weight of the final product multiplies in the same order. Hence, the properties of the final polymer product are different from the parent monomer.

In a commercial operation, the degree of polymerization is controlled in order to produce the polymer of desired properties. For example, in the plastic industry, the degree of polymerization is controlled to produce low-density polymer or high-density polymer, depending on the quality of the plastic desired.

The polymerization reaction is thermodynamically favorable at lower temperatures, as it is an exothermic reaction. The energy released is based on the bond energies of the monomer and in an ideal case is equivalent to the difference in C-C bond energy and C=C bond energy of the molecule.

During the polymerization reaction, the number of moles decreases as the degree of polymerization increases; hence, the reaction is favored with the increase in pressure.

Polymerization reaction can also be controlled with the reaction time. At a steady state, the polymerization reaction proceeds at a constant rate over time, with a decrease in monomer concentrations.

In a thermal process, the polymerization reaction can be explained through radicalic reaction mechanism as described in Chapter 2.2, through termination reaction in step 3.

However, modern commercial polymerization units are based on catalytic processes, which essentially accelerate the process at a lower temperature with a higher rate of production.

References for Chapter 4

1. Ullm ann's Encyclopedia of Industrial Chemistry, Wiley-VCH, Weinheim (2000).
2. L. A. Kohl and C. F. Riesenfield , *Gas Purification* , McGraw Hill, New York, (1960).

5

Shale gas (Tight gas) / Shale oil (Tight oil) development

5.1 The definition

- Shale is a rock formation created by the deposit of sediments from marine water, mostly made of clay and silt. Hydrocarbons, which are basically similar to natural gas and petroleum oil, are trapped in the solid rock. The permeability of the rock is close to zero (<0.1 mD (milli-Darcy), i.e., the hydrocarbons are trapped so tightly that they are practically immobile and the oil contained in these reservoirs will not flow to the well bore.
- The Canadian Society for Unconventional Resources defines the hydrocarbons produced from those rocks as "tight oil." Usually, the lower hydrocarbons, which are in the gaseous state at normal atmospheric conditions, are called "shale gas/tight gas" and the liquid hydrocarbons are called "shale oil/tight oil."
- The difficulty in the recovery or extraction of those "tight" hydrocarbons led to the development of a new process, which is a combination of two technologies developed earlier separately: (i) horizontal drilling and (ii) multi-stage hydraulic fracturing. The process is well known as "fracking."
- Just to remind readers here not to confuse "shale oil" with "oil shale"—they are not the same material. Oil shale is basically solid organic matter, created in early stages of development over millions of years ago, known as kerogen. Kerogen is not even an oil, and the shale is not even a shale, another misnomer in the oil industry. In order to convert kerogen into oil, it needs to be retorted at a very high temperature (>500 °C) in a rotating furnace.

5.2 Background

The shale revolution has changed the game for the North American energy industry over the past decade. In 2008, the US was believed to be in short supply of natural gas resources and in need of LNG imports. Now, a decade later, the region has transitioned to a gas exporter, with LNG import terminals converting

Table 5.1. Major tight oil and tight gas production areas in decreasing order of production (Source: EIA)

Tight Oil Production areas	Tight Gas Production areas
Eagle Ford (TX)	Marcellus (PA, WV)
Bakken (MT, ND)	Haynesville (LA, TX)
Three Fork (ND)	
Permian (Wolfcamp)	Eagle Ford (TX)
Permian (Bonespring)	Fayetteville (AR)
Permian (Spraberry)	Bamett (TX)
Permian (Delaware)	Woodford (OK)
Haynesville (LA, TX)	Bakken (ND)
Utica (OH, PA)	Utica (OH, PA & WV)
Marcellus (PA, WV)	
Woodford (OK)	
Grante Wash (OK)	
Austin Chalk (LA)	
Monterey (CA)	

5.3 Production

As already mentioned above, the shale industry growth has been possible due to the development of new technology—fracking—which injects high pressure water along with sand and chemicals down into the ground to fracture the shale rock, in order to force out the oil and natural gas trapped in those layers of rocks. Fracking technologies had been around for decades and were used in vertical wells in the past. However, the breakthrough came from deploying vertical-horizontal wells and fracking in combination called "hydraulic fracturing." The term "tight oils or tight gas" is derived from the fact that the oil and gas deposits are tightly held within geological formations and are not free flowing, as the rock is very dense and not porous, i.e., poor permeability (see Chapter 1, Figure 1.5). That means the oil and gaseous hydrocarbons are trapped so tightly that they are not mobile even within the rock. So the only way the hydrocarbons can escape or be recovered from the rocks is by artificially cracking the rocks and opening up the pores.

That is the reason why fracturing is needed in order to increase the permeability of the rocks (Figure 5.2). It is called hydraulic fracturing because high speed and high pressure water jets are used to fracture the rocks. Once the pore is created, the trapped hydrocarbon is released and simply by pressure differences it reaches the surface. But the pores do not remain open for a long time and the production decreases. The closure of the fissure is prevented by introducing fine solid sand particles called "proppant" into the formation. In the beginning, simply sand particles were used, but now synthetic nano-particles are under investigation for the better performance. Production of tight oils would not be economically viable without these technologies.

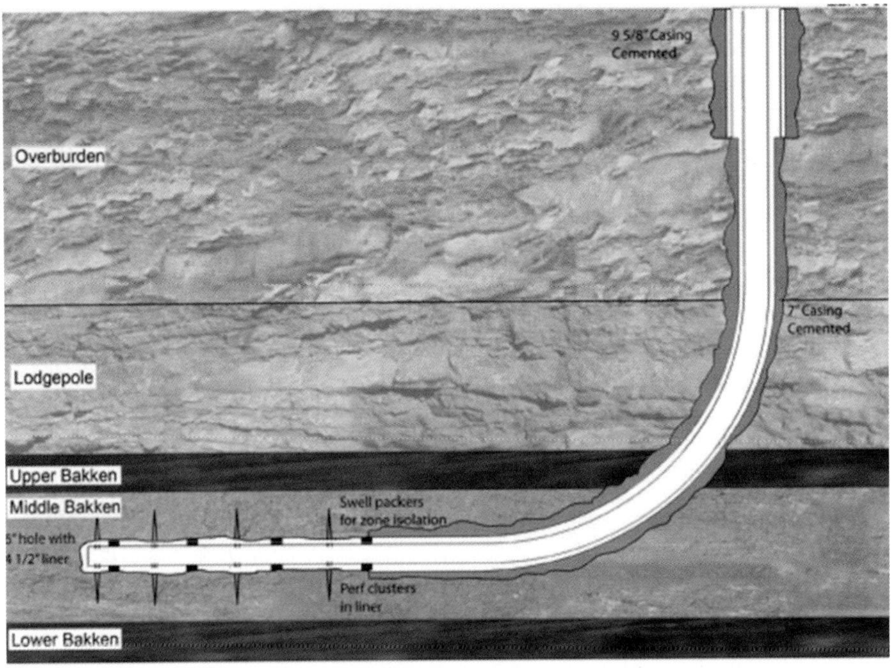

Fig. 5.2. Different subsurface environments in which hydraulic fracturing takes place. (Fig. source: Energy and Environmental Research Center, University of North Dakota)

According to a report published (2014) from the "Office of Water" of the EPA, additives are mixtures of chemicals (may vary up to 30) that are added to the base fluid, which is water, to change the properties of the fluid in order to enhance the production of the tight oil. "Produced water" is defined as the water that flows from and through wells to the surface. "Waste water" is the produced water that is being disposed of above ground by various practices. The current fluid disposal methods can be costly due to the transportation to the sites, and these fluids are often simply pumped into a salt water disposal well where they are permanently settled. Current treatments of these fluids are often limited due to costs and efficiency.

As the fluid is injected inside the well, pressure increases until the rock formation fractures. Maximum pressure applied to the wells may vary and is reported to be somewhere between 2,000 psig to 12,000 psig (14 to 80 MPa.). During the same period, a well experiences temperature change too. A well can experience multiple cycles if re-fracturing is done.

Initially, oil is produced by liquid expansion across the pressure drop from the reservoir to the well bore. Later, the gas dissolved in the oil expands and lifts the oil to the surface.

While it sounds very attractive in the short term, there are longer-term threats as well. As shown in Figures 5.3 and 5.4, the shale wells' production increases rapidly for the first few months and then rapidly declines. In fewer than 24 months, the production decreases to a low level such that it is no longer economical to continue the process. Thus, in order to increase production, more and more wells are needed every month. Just for the sake of comparison with conventional reservoirs of light oil such as in the Middle East, the production profiles remain constant for a longer period of time, and one can estimate how much oil is expected to be produced in the near future.

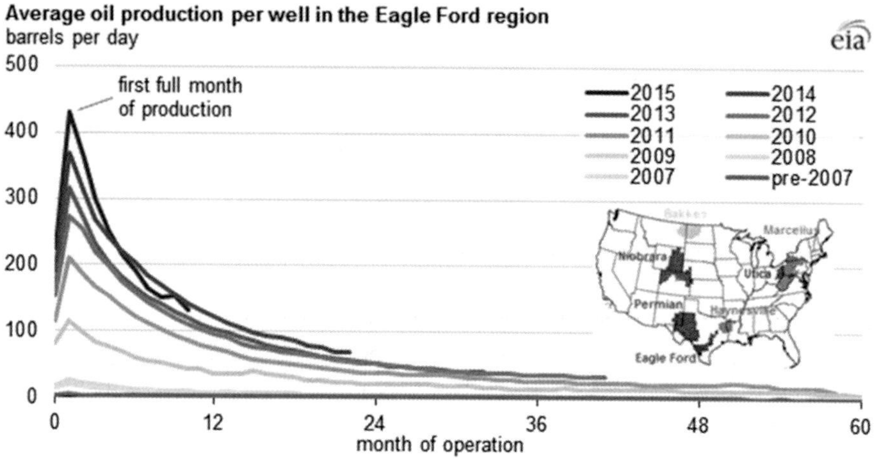

Fig. 5.3. Eagle Ford oil production over time (source: EIA)

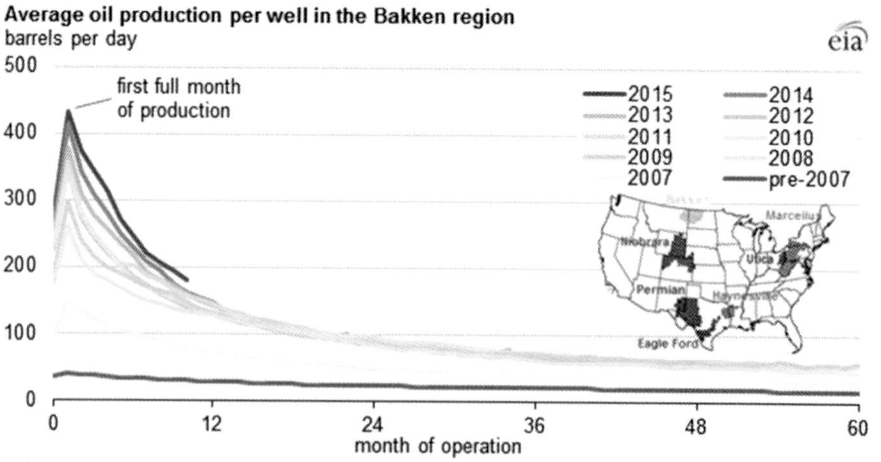

Fig. 5.4. Bakken oil production over time (Source EIA)

Increasing production of shale gas in the US will have a great impact on the petrochemical industry. Because of the availability of cheap and abundant natural gas, ethylene and propylene industries are now switching to ethane and propane feedstocks instead of liquid hydrocarbons. This in turn increases the demand for ethane and propane. At present, gaseous hydrocarbons are less than one fourth the price of liquid hydrocarbons in terms of energy value. As demand increases, the feedstock cost advantage may disappear in the near future.

According to a report by the Gas Processing and LNG group (Oct 2018), the North American gas demand will grow at a rate of 2% per year due to the increase in US LNG exports and the retirement of coal power plants. Rapid development of the shale industry could mean a large increase in investment.

North America has major advantages when it comes to consistently producing plentiful and low-cost natural gas. Technological improvements and increased efficiency have driven down break even prices for natural gas production in North America, making areas like Marcellus and Utica more profitable. According to the IEA report (2018), North American gas prices are expected to stay below $3/MMBtu for decades to come. The shale gas boom established the US as the world's leading natural gas producer and is responsible for billions of dollars of investment in the US gas processing industry. As LNG is becoming increasingly important to global gas markets, the US has undergone extensive LNG terminal reconfiguration and construction to capitalize on the opportunity.

5.4 Properties

The value of a crude oil depends on the API of the crude and the volume and the quality of the major fractions such as naphtha, distillate, and gas oil. When shale hydrocarbon is recovered, it usually contains condensate which is an ultra-light liquid fraction mainly consisting of hydrocarbons below C_6 of API above 50 degrees. Though condensate is part of the crude oil, it is usually sold separately as a solvent or diluent and separated from the major tight oil portion for transportation. Figure 5.5 (source: EIA report 2017) illustrates the variation of API of the condensate from the Eagle Ford area and the corresponding tight oils from Bakken and Eagle Ford over time. Figure 5.5 indicates that the API of the condensate varies between 50 and 55 degrees and that of tight oil between 40 and 45 degrees over time.

Due to the increasing production of tight oil, the condensate has flooded the market in recent years. That is the reason, though they are valuable hydrocarbons, they are sold at a discounted price as compared to the light tight oil. Excess of these materials is actually a problem for refiners to handle, especially in the crude distillation unit.

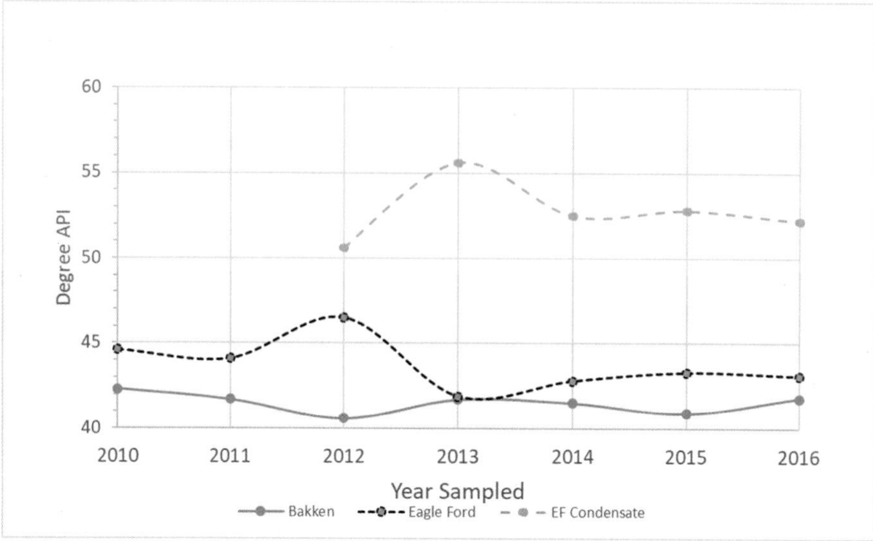

Fig. 5.5. Variation of API with the year of tight oil and condensate production

During the same period, the sulfur content in the Bakken crude varied only between 0.10 to 0.15 wt% and that of Eagle Ford crude between 0.10 to 0.20 wt%, whereas the sulfur contents in the condensate are negligible and may be in most cases dissolved hydrogen sulfide reported as sulfur.

Data in Table 5.2 show that the light tight oils (LTO) are usually a little bit lighter than WTI and consist of sulfur a little less than WTI. The assay shows that tight oil will produce lighter fractions (naphtha + distillate) almost the same as WTI, and the amounts of heavier gas oil and resid lie within $\underline{+}1\%$ point. Thus, a refinery handling WTI may be able to process the light tight oil easily though there are some other problems discussed later.

The chemical properties of tight oil vary from one region to another, as for example, between Bakken and Eagle Ford crudes. In most cases, Eagle Ford LTO has lower sulfur and nitrogen than that of the Bakken LTO. Asphaltenes content is almost negligible in both the LTOs but Bakken contains higher resid and less wax as compared to that of the Eagle Ford LTO.

The gas oil content in the light tight oil from various regions varies between 20 to 25 wt% while the average resid (525 °C + / 100O °F +) content is about 5 wt%. Thus, a deep conversion refinery processing heavier crudes will not be able to handle the tight crude oil.

Table 5.2. Comparison of the average properties of a typical light tight oil with marker crude WTI (**Source:** U.S. Energy Information Administration, Short-Term Energy Outlook, January 2017)

Properties	Temp range, °F	Light Tight Oil (average)	WTI
API Gravity		>42	42
Sulfur, wt%		<0.25	0.35
Naphtha, wt%	100-330	30	32
Distillate, wt%	330-650	40	38
Gas oil, wt%	650-1000	25	23
Resid, wt%	1000F +	<5	7

All the wax content in the tight oil falls mostly in the gas oil range or is reported as resid hydrocarbons. The average properties of the paraffinic wax present in the tight oil are as follows:

- Carbon number > 30
- MP ~ 40 °C/100 °F
- BP ~ 370 °C / 700 °F
- Density about 900 kg / m^3

Wax is not a desirable component as it is responsible for fouling. Pour point and viscosity both are impacted by the wax content, and it restricts oil movements in the tank and in the pipeline to the crude unit. Dewaxing is usually done with ketone solvent and crystallization. It could be done before going through the crude tower and avoid furnace fouling.

Filterable solids are fine inorganic particles that comes from the fracking of the rocks. These are difficult to separate from the organic portions. Eagle Ford has much higher filterable solid as compared to Bakken. It is not a desirable component as it enhances the fouling tendency of the tight oil inside the furnace wall.

5.5 Unconventional crude of conventional nature

The shale hydrocarbons are trapped and so tightly bound inside the rocks that they cannot escape and there is no need of a cap rock such as in conventional oil or gas reservoirs. Thus, new unconventional technologies are required to develop the shale reservoirs.

As already discussed, the light tight oil is recovered by unconventional means using fracking technology. Thus, the LTO becomes an unconventional crude; however, the carbon number distribution of the LTO shown in Figure 5.6 indicates that the LTO has properties quite close to a conventional light crude oil. The results were determined from the simulated distillation (SimDist) curve following ASTM D-2887 specification.

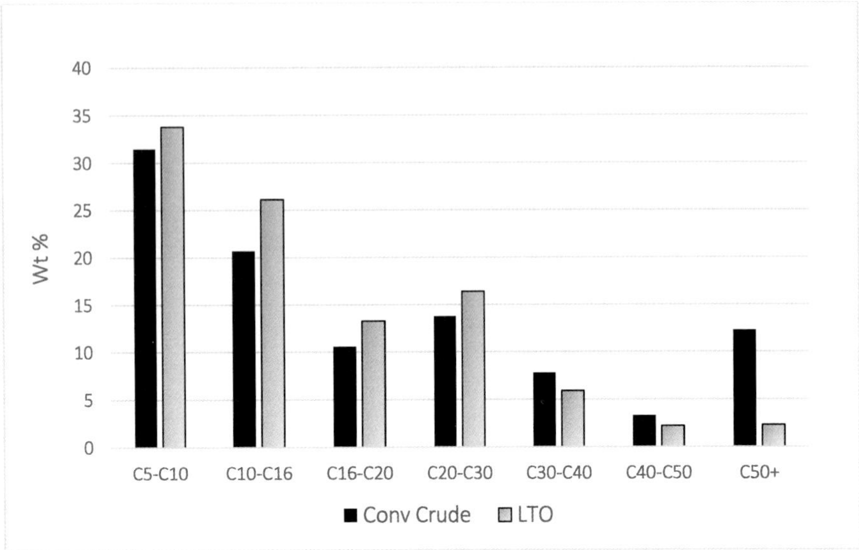

Fig. 5.6. Comparison of the carbon number distribution of LTO of API 42 with conventional crude of API 32

Except for the C50+ resid fractions which are present at very low concentrations (<5%) in the LTO as compared to conventional crude oil (>10%), all other carbon distributions are almost of the same order, and the heavier the molecular weight, the lower the concentrations. Nevertheless, results also illustrate that all the hydrocarbons of carbon number less than C30 are at a higher concentration at least by 2 to 3 wt% in LTO (which is also lighter crude) as compared to conventional crude. Thus, the thermal behavior of the distillate fractions of the LTO is expected to be almost the same as that of conventional crude during the refining processes.

However, it is also to be noted here that the resid fraction in conventional crude is mostly aromatic in nature and that of the LTO is paraffinic and waxy in nature. That changes the coking tendency of the resid fractions during the thermal cracking process of the LTO resid as compared to the conventional resid.

5.6 Impact of LTO on refinery operations

As shown in Table 5.2, tight oil is a sweet light crude with very low sulfur and is thus very attractive for refiners, as it takes little to refine it. Low sulfur crude helps across the refinery as it reduces sulfur load especially in the hydrotreating and hydrocracking units, which means lower hydrogen consumption and longer catalyst life. It has a great effect on the environment as well, as it reduces H_2S

formation and decreases SOx / NOx emissions. It can easily meet the stringent sulfur specifications in gasoline and diesel transportation fuels. But it can also create a number of refinery processing challenges discussed below.

As soon as tight oil entered the oil industry, it impacted not only the refineries but also the pipeline industry. Transportation of tight oil as a blend to the refinery results in problems for the pipeline industry. Blending light tight oil with heavier crudes to make it a crude that a refinery can handle makes it a dumbbell shape crude with properties of two ends of the spectrum. That also creates compatibility issues. High paraffin content of tight oils greatly increases the potential impact of asphaltene precipitation upon blending with conventional crudes and especially with heavy oil. High wax content in the LTO is also a transportation issue besides refinery fouling issues.

Additional challenges are faced across the refinery as tight oil may contain:

- chemicals from the base fluid
- stabilized emulsions in the de-salter
- increased potential for system corrosion and fouling
- negative impact on waste water treatment

Tight oil characteristics can vary greatly from batch to batch, even within the same type of crude oil supply:

- API gravity for tight oils can vary from 40-55° depending on the amount of condensate present.
- filterable solids can be higher than the conventional crude oils, with greater volume (>500 ppm) and smaller particle size passing through the de-salter.

Another concern with tight oil handling is the potential for low temperature waxes to form. These waxes contain long-chain paraffins that can be problematic because of:

- Poor cold-flow properties
- Poor de-salter performance
- Impact on storage tank capacity through settlement
- Increased potential for fouling in the cold train

It is increasingly important for refiners to be able to identify, interpret, and respond quickly to changes in crude feed properties. Blending of tight oils into the standard crude slate may make normal refinery operations difficult to maintain. Processing these blends can have a significant negative impact on overall profitability, affecting product quality, unit reliability, and on-stream time. Determining how a new crude oil fits into a refinery operation requires a detailed understanding of the physical and chemical characteristics of that crude.

There are several potential side effects when a new crude is introduced to a refinery. The damage depends on refinery configuration. The further such a crude departs from the refinery design, the greater the likelihood there will be adverse side effects. This is why usually refiners run new crudes as a blend to mitigate the risk. While there is no perfect method for determining future processing risks with new crudes, several key parameters should be examined while processing to help determine risks: API gravity, sulphur and nitrogen contents, acid number (TAN), pour point, viscosity, metals, salts, etc.

Some of the common issues a refinery may face when shale oil is included in the recipe:

- Increased salts contaminants contribute to fouling in hot furnaces
- FCC units face heat balance issues due to lack of coke formation
- Delayed coker plant suffers from lack of feedstock availability
- Coke drum excessive foaming, hot spots, coke cave-ins
- Longer cycle times due to lower coke made in the delayed coker
- Calcium and iron present in tight oil poisons catalyst in various downstream hydrogenation units
- Bottlenecks with lots of light ends flashing
- Corrosion increase in overhead systems

5.6.1 Impact of tight oil on the FCC unit

Fluid Catalytic Cracking (FCC), a high temperature thermal cracking process (though in the presence of a catalyst), is considered the money-printing machine for a refinery. It is used for the conversion of heavy gas oil into lower molecular weight gasoline range hydrocarbons. Processing only tight oil in a refinery impacts an FCC unit a lot, because of large variation of the gas oil content in the tight oil, depending on the source. In some cases, low VGO content leads to shortage of the feedstock for the FCC unit, and running it at a lower capacity makes it less economical for the refinery. In order to run the FCC unit at full capacity, refiners need to purchase fresh VGO from a different source, and that may have different qualities as compared to tight oil VGO. For example, more paraffinic tight oil VGO yields higher saturated gasoline of lower octane value as compared to conventional VGO. Another problem with this VGO is that it yields very low coke as compared to the conventional VGO and hence there is a problem of maintaining heat balance in the regenerator. The regenerator temperature (around 1200 °F) is maintained by burning the coke, and low yield of coke provides less heat. Refineries need to purchase additional heat sources in order to maintain a stable heat balance. Blending with a resid feed to maintain heat is not an easy solution as the catalyst for the resid treatment is of different qualities than that of the normal VGO. Switching catalyst impacts the profitability of the refinery.

Tight oil contains a relatively large quantity of metals or salts. The most important players are iron, sodium, and calcium. Metal deposit on the catalyst bed poisons the catalyst, keeping the unit performance low.

According to a report published by Grace in Grace Catalyst Technology (2014) to show the performance of their catalyst, they reported that when Bakken whole crude was processed through their FCC pilot plant at 980 °F, it produced 14.6% LPG, 65.4% gasoline (C5-430 F), 14.2% LCO (430-700 F), and 2.3% 700F+ bottoms with 1.8% coke formation over the catalyst surface. When the whole crude was replaced by 650F+ bottoms, which is actually a regular feed for an FCC commercial plant, the gasoline production decreased by about 12% with coke production doubled.

5.6.2 Impact of tight oil on the delayed coker plant

As it has been seen that tight oil composition varies a lot from well to well, however, the amount of resid always remains low, almost below 5% of the crude, at the expense of higher gas oil or distillate yields. Another big concern for refiners is that most of the higher boiling materials are paraffinic wax. Thus, the overall resid fraction is low in aromatics content with very low asphaltenes and low in carbon residue (CCR), and those are the factors responsible for promoting coke formation. Thus, utilizing tight oil in a delayed coker not only makes it run at a lower capacity, but also it lowers the coke production, and that will increase the run length of the coker drum, decreasing the economic benefit. Most refiners with large coker capacity run a blend of tight oil with heavy crude, but that may involve other problems such as compatibility issues and increased fouling.

5.7 Thermal processing of light tight oil

There is little doubt that LTO is drawing great attention in the energy industry as it is adding to the economy. LTO being in the market for a short period of time, very few data are publicly available on the thermal processing of LTO. There is potentially great interest for refiners as well as for the petrochemical industry at large to have some new data on the thermal behavior of LTO as a whole crude.

However, there is a problem associated with the thermal processing of LTO. In the oil industry, the "conversion" makes a lot of sense; it has technical as well as economic significance. As discussed in earlier chapters, the lower the molecular weight of the hydrocarbon feedstock, the higher the temperature it requires to carry out pyrolysis or thermal cracking. The conversion is a combination of residence time and temperature, in other words, the "severity." As shown in Figure 5.6, in the carbons distribution of LTO crude, almost half of the crude oil is below C16-range hydrocarbons, and they are basically smaller-sized molecules. Usually for these materials, the thermal processing requires more than 900 °F temperature. Almost 95% of the hydrocarbons in LTO are volatile at that temperature, thus having a very short residence time of each component in the crude oil to undergo any significant conversion.

The present author has generated some important results of significant value on thermal processing of LTO for the industry. Two sets of results were produced at 930 °F:

(i) the whole LTO crude and (ii) (>C30) bottoms of the same LTO.

In the same experiment, almost double the amount of C_1–C_4 gaseous hydrocarbons were produced from the heavier bottoms feedstocks (4%), as compared to the lighter whole crude (2%). As explained in earlier Chapters 1 and 2, it is easier to break C-C bonds or C-S bonds for a longer chain hydrocarbon as compared to the smaller hydrocarbons and hence the heavier bottoms feed (> C30 range) produced more gaseous hydrocarbons.

On the other hand, the heavier bottoms fraction has bigger molecules containing some aromatics and CCR (Conradson carbon residue) that promote coking reactions and hence produced a larger quantity of coke (8%), as compared to the lower molecular weight lighter feedstock (1%).

The whole crude yields over 97% liquid, confirming the fact that the lower molecular weight paraffinic hydrocarbons did not undergo enough cracking and remained within the liquid range. The main reason for this is that the majority of the fraction (< 900 F) is in the vapor phase under thermal reaction conditions, and there was not enough time for significant conversion. For the same reason, there was not enough coke (below 1%) formation, too, from the whole crude.

Table 5.7 shows results with special emphasis on the production of C_2 to C_4 olefins for the petrochemical industries. Some interesting results on olefin production are obtained mainly indicating the domination of dehydrogenation reactions of corresponding paraffins over cracking reactions. However, the total amount of gaseous hydrocarbon production is lower (1.0%) from the lighter whole crude feedstocks, but higher yields of olefins were observed. The olefin to paraffin ratios for all three hydrocarbons (C2, C3, and C4) are more than double that from the lighter feedstock, as compared to the heavier bottoms feedstock. The olefin to paraffin ratio also increases with the increase in molecular weight from C2 to C4 hydrocarbons. Actually, in the case of whole crude, the ratio almost doubled from C2 to C3 and again from C3 to C4 hydrocarbons, which clearly shows that dehydrogenation is more favorable with the increase in molecular weight of the paraffin hydrocarbons. As discussed in earlier chapters, that all thermal reactions taking place through intermediate free radical formation, and in the case of the lighter paraffinic hydrocarbons, once the free radical is formed, dehydrogenation was favored in the vapor phase over further cracking reactions. Consequently, in the case of the heavier feedstocks, multi-steps are involved, and the cracking reaction is favored for the bigger molecules first, followed by dehydrogenation of the cracked free radicals, before the production of lower olefins—thus, within a shorter residence time in the vapor phase, the yields of the dehydrogenation products are low.

Table 5.7. C2–C4 Hydrocarbons production in kg/ ton feed - using LTO bottoms and whole LTO crude feedstocks at 930F

Product	Yields from 850F + Bottoms			Yields from whole crude		
	Paraffin	Olefins	Olefin / Paraffin ratio	Paraffin	Olefins	Olefin / Paraffin ratio
Methane	9.224	x		3.947	x	
C2 Hydrocarbons	7.0122	2.021	0.288	3.551	2.151	0.606
C3 Hydrocarbons	6.799	4.944	0.727	2.579	3.310	1.283
C4 Hydrocarbons	4.557	4.502	0.988	0.744	1.817	2.442
Hydrogen			0.177 kg/ton			0.118 kg/ton

Due to the high paraffinic nature of LTO and the domination of the dehydrogenation reaction, Table 5.7 also shows that there is some hydrogen production (between 0.1 to 0.2 kg /ton feed) during this process, which can be utilized in a refinery as a byproduct for another plant.

One can process less expensive LTO feed abundant in the US, either straight or in blend. Of course, with the additional cost of adding super-heated steam, the result will improve further at the optimum conditions of a commercial steam cracker plant.

5.7.1 Impact of thermal processing of LTO on the product quality

At higher temperatures, inter-conversion of various hydrocarbon species takes place randomly. Concentrations of active species changes as the reaction proceeds, with the formation of active free radicals and the capping of those with other free radicals or hydrocarbons, which have a profound effect on the products yields. Most of the LTO (>90%) is in the vapor phase under the reaction conditions, and the most reactive plausible reactions in the vapor phase are through isomerization, dehydrogenation, cyclization, and polymerization reactions over cracking reactions. That is the reason when the whole crude was processed at 930 °F, less than 2% lower hydrocarbons (<C4) were produced, with about 97% liquid recovery, with no additional production of lighter naphtha (C5–C10) fraction.

However, the most significant difference noticeable in the liquid product was that almost all of the C50+ heavy waxy resid material had been destroyed and was converted into smaller molecules (<C40) and very little (<1%) coke. There were small changes of hydrocarbons in each group, mostly due to the molecular shift and not necessarily due to the cracking within the groups.

Nevertheless, one concludes that it is difficult for the heavier molecules to get converted into naphtha-range hydrocarbons through cracking under these conditions. On the other hand, a slight increase in the middle-range hydrocarbons indicated that the bigger molecules (>C40) cracked near the middle and produced molecules of near C20-range liquid hydrocarbons instead of smaller molecules (<C10). This is the reason why, in spite of thermal cracking of the heavier

molecules, the overall liquid yield remained almost the same with less than 2% gaseous hydrocarbons.

5.8 Impact of tight gas / tight oil on the petrochemical industries

Tight oil essentially can be used as transportation fuel because a large volume of gasoline and diesel fractions are present in the LTO. But tight gas has the biggest impact on the petrochemical industries. In the US, ethane is the primary source for the production of ethylene, and ethane primarily coming from the tight gas/shale gas. All forecasts say there will be constant increase in ethylene demand in the near future, and all future ethylene plants are going to use pure ethane as feedstock (O&GJ June / July 2018).

Tight oil is a light sweet crude of better quality than that of the Middle Eastern or Western African light sweet crudes. It is now available abundantly in the US and at a lower price. As shown in Figure 5.8.1, it is a boom for the petrochemical industries as they have several options to choose from, utilizing lower-priced feedstocks for the production of olefins (see for details Chapter 6). Beside this, they can also produce hydrogen as a byproduct, while producing additional hydrogen as well using tight gas as feedstock for the steam reformer and selling it to the refineries.

Thus, it looks like a golden age for the refineries as they get low-priced feedstock and produce high-demand product for the petrochemical industries. Until the shale boom, the natural gas price was all over the map, fluctuating widely, and no one could predict the future. Since the development of fracking technology, the cost of production of shale oil and shale gas is continuously decreasing, and now the cost of natural gas has been almost stable for some time and is expected to remain so in the near future. But the cheap gas price may create another problem for the gas industry. Gas demand will increase—there is already a shift of natural gas into power generation plants, replacing coal—and there may be some replacement of liquid hydrocarbons by gaseous hydrocarbons in other industries such as transportation. In the end, the price of natural gas will increase and hence will impact all other industries.

At present, the production of tight oil and tight gas is booming in the United States and Canada, but it will be quite difficult to maintain in that way because the production per well decreases rapidly with time (see Figures 5.3 and 5.4). In order to keep up with the same production range of the hydrocarbons, drilling activities need to be increased at any cost.

Due to the excess availability of natural gas from the shale fracking industry in the US, it is worthwhile to discuss a typical scenario for the shale gas industry to produce on-site olefins for the petrochemical industry. Figure 5.8.2 represents a most probable wholly integrated system, a tight gas/tight oil production site, which is further connected to a fractionator, steam cracker, and olefin separation plants.

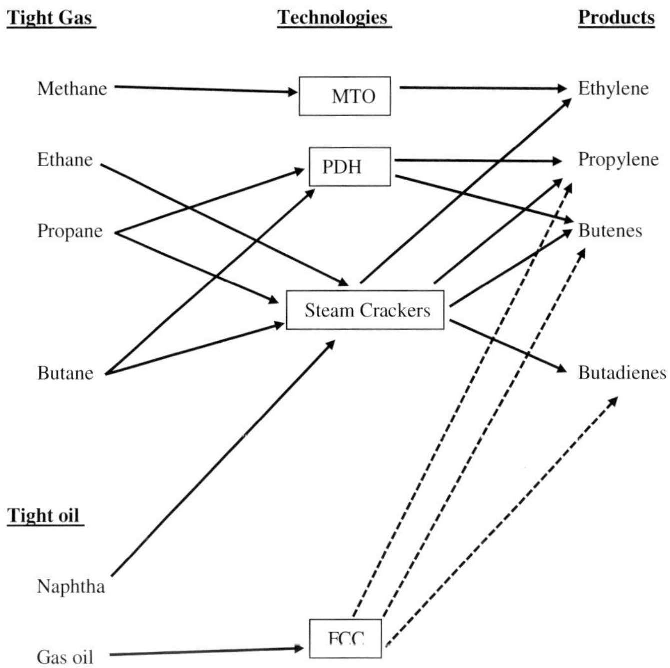

Fig. 5.8.1. Various options of tight gas/tight oil for the petrochemical industries

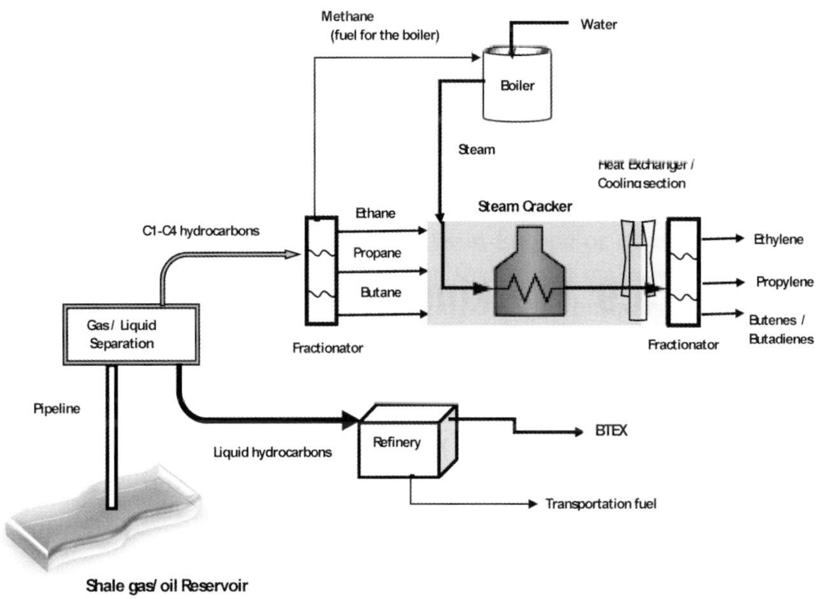

Fig. 5.8.2. A typical proposed schematic diagram of a petrochemical complex for the shale gas/oil industry

6

Petroleum to petrochemicals

6.1 Basic definition

Petrochemicals are the hydrocarbon materials derived from crude oil that are used for the manufacture of specific organic compounds other than the fuel. The most common petrochemicals are "olefins" and "aromatics," which are used for the manufacture of plastics, solvents, or chemicals. The most high-value olefins are ethylene, propylene, butenes, iso-butenes, butadienes, and iso-prenes. The corresponding high-value aromatics are benzene, toluene, ethylbenzene, xylene (BTEX), and styrene.

6.2 Introduction

Things are moving so fast that by the time this book is published, a lot of things may have changed. In this digital age, things are moving at the speed of light. The purpose of this sub-chapter is to make the readers aware of the situation and prepared to face the future.

According to the recent forecasts released by several energy news outlets[1], natural gas will take over as the main fossil fuel, over solid or liquid hydrocarbons, by 2040. The US is expected to dominate the energy market beyond 2020 with increased tight oil/tight gas production and a secure supply of natural gas to satisfy global demand. Given the impact of climate change, we are heading towards a global energy transition with natural gas playing an important role. Whether it is in the form of natural gas, or compressed gas (as CNG) or liquefied gas (as LNG), global cooperation and partnerships are attracting international investments[1].

Nevertheless, market demand and economics dictate the pathway the industry would like to follow. Refineries are now looking for an alternative as they are facing a lower profit margin in transportation fuel. Due to the surge in polymer consumer demand, important global players are now focusing on petrochemicals more than ever to narrow the gap or shortage of plastics in the expanding electronics industry.

As per several publications in the O&GJ during the summer of 2018, petrochemicals are becoming the fastest growing sector in the petroleum industry. They see a significant increase in investment for future development of the downstream industry in the Middle East. The name of the game is now "digital transformation," where the focus is customer-oriented—identifying and exploring the opportunities and providing support. Saudi Aramco is leading the way by introducing a game-changing project called "Crude-oil-to-chemicals," implemented jointly with SABIC. It is a massive undertaking, using first-of-its-kind technology developed internally, resulting in a product that is less expensive to produce with a much lower carbon footprint.

O&GJ (October 2018) further confirmed that Saudi Aramco and Total SA have signed an agreement to launch a front-end engineering design (FEED) for their previously announced proposal to add an integrated petrochemical complex downstream of their jointly held Saudi Aramco Total Refinery & Petrochemicals Co.'s (Satorp) 440,000-b/d full conversion refinery in Jubail, Saudi Arabia. The new complex will include a mixed-feed steam cracker—50% ethane and refinery off gas—with a capacity to produce 1.5 million tons/year (MMTPA) of ethylene and related petrochemical units designed to yield an overall production of more than 2.7 MMTPA of high-quality chemical products.

Next in line is Oman Oil Refineries and Petroleum Industries Company (Orpic), aiming to invest in another massive $4.5 billion-dollar Liwa Plastics Industries Complex (LPIC) project to diversify their portfolio of vast natural gas resources. They are building a new polyolefin plant to convert gaseous hydrocarbons into olefins for the plastic industry.

In another part of Asia, Reliance Industries in Gujarat, India, is investing in liquid hydrocarbons. They are the major producers of ortho- and para-xylenes. Ortho-xylene is the feedstock to produce primary material for paints and fiberglass, whereas para-xylene is the feedstock for terephthalic acid used for the manufacture of polyester fibers. Reliance is also investing in a highly integrated 1.5 MMTPA ethylene cracker plant in its own refinery to produce feedstocks for polyethylene polymers.

On the other hand, as per China's state oil company news, China is also investing in a 0.6 MMTPA plant of solid fossil fuel coal (other than natural gas), utilizing clean coal technology based on gasification. In this process, coal is gasified to syngas, a mixture of carbon monoxide and hydrogen, which is further converted into monoethylene glycol, a petrochemical feedstock (see Chapter 6.6).

In other parts of Far East Asia, namely Vietnam and South Korea, it is also planned to expand the petrochemical business by investing in integrating refining and the petrochemical business. According to a report in the O&GJ (Aug 2018), Vietnam has already awarded a contract to build an ethylene plant using a mixed-feed flexible cracker, and South Korea is carrying out a feasibility study for the proposed construction of a 1.5 MMTPA steam cracker olefin plant in the Ulsan refinery. Downstream olefin facilities will produce polyethylene and polypropylene plants to meet Asia's rising demand for plastic industries.

O&GJ (July 2018) reported that ExxonMobil Chemical Co. has started up its 1.5 MMTPA ethane steam cracker at the company's integrated chemical and refining complex in Baytown, Texas. The new cracker will provide ethylene feedstock for the 650,000-tpy high-performance polyethylene plastics plant in Texas. This new ethane cracker will help meet the growing global demand for high-performance plastic products.

It is interesting to observe that the demand for ethylene increases with the increase in the global economy. According to publicly available information in an Argus Ethylene 2018 report, the actual demand for ethylene in the US increased to annual growth of 2.8%, more than the forecast due to a better economy, and it is expected to grow to about 4% for the next few years.

However, it is a challenge for the refiners to meet the demand with reasonable economic benefit. Economic evaluation of the production of olefins is highly dependent on the site-specific projects.

The most dependable factors are:

(i) availability of natural gas
(ii) cost of the feedstocks
(iii) cost of installation of a pyrolysis unit and steam cracker unit
(iv) separation of the byproducts, or purity of the final product
(v) cost of converting olefins into polymers
(vi) market demand and the cost of transportation of the feed and products

Finally, it depends on whether the final product is consumed within the same geographic area or at a distance. For example, for the US to reach the global market, the shipping costs are about 12% more as compared to the market within Asia[1].

On one hand, the availability of low-cost hydrocarbons from tight gas production has increased investment in steam cracker plants in the US. On the other hand, further East, they are concentrating on integrated steam cracker and refinery plants, with liquid hydrocarbons as feedstock.

The demand for propylene has also increased over the years and is growing at a faster rate than ethylene. Propylene is used in a wide variety of petrochemicals, such as propylene oxide, acrylonitrile, cumene, acrylic acid, etc. Polypropylene is used in the polymer industry, and its use as plastic has increased, especially in the automobile industry.

According to an International Energy Agency (IEA) forecast published in October 2018, future oil demand will increase, not due to the transportation industry but rather due to the increase in petrochemicals demand. There are wide varieties of petrochemicals such as fertilizers, plastic packaging materials, clothing, digital devices, medical equipment, detergents, and tires—besides the demand in the area of cars and airplanes. Though there is increasing interest in moving towards renewable energy, this actually creates additional demand for petrochemicals; renewable energy equipment, like solar panels, wind mill turbines, insulators, electric vehicles, batteries, etc., require plastic parts. According to the IEA report, after decades of decline, the United States has emerged as the low-cost petrochemical production place, mostly due to the tight oil revolution, and 40% of the global ethane-based petrochemical industries are now in the US. However, there are also environmental challenges with the increasing demand for the use of plastics.

6.3 Integration of refinery and petrochemical complex

Refineries are always looking for ways to increase their revenue. That's why existing refineries are looking for opportunities for integration of refinery and petrochemical industries. As discussed above in the introduction, new refineries are already building grassroots integrated complexes around the world.

However, refiners are facing increasing challenges with the increase in stringent environmental regulations. There are constant changes regarding feedstock qualities and quantities. More than ever before, refiners are facing cut-throat competition. The only way they can survive is through broadening their portfolios, maximizing their profits through process integrations, implementing new and advanced technologies, and digitizing their operations. They must maximize their target product production through multiple sources and multiple feedstocks within their refinery. Flexibility is the "name of the game" for survival in the current situation.

Figure 6.3 illustrates a snapshot of various pathways and opportunities, from the initial paraffinic feedstock in the refinery to producing olefins for the petrochemical industries, and then the various options for final plastic or polymer end-use utilization.

There are several advantages to multi-integrated plants:

i. Most importantly, a capital cost advantage—by eliminating some common units such as fractionations or furnace tubing
ii. Feedstock transportation cost—transferring feedstocks over the fence
iii. Availability of byproducts from one unit to be utilized in another plant—such as fuel gas or hydrogen
iv. Reduction of energy utilization through integration—improved efficiency
v. Manpower reduction—possible elimination of duplication of the work force

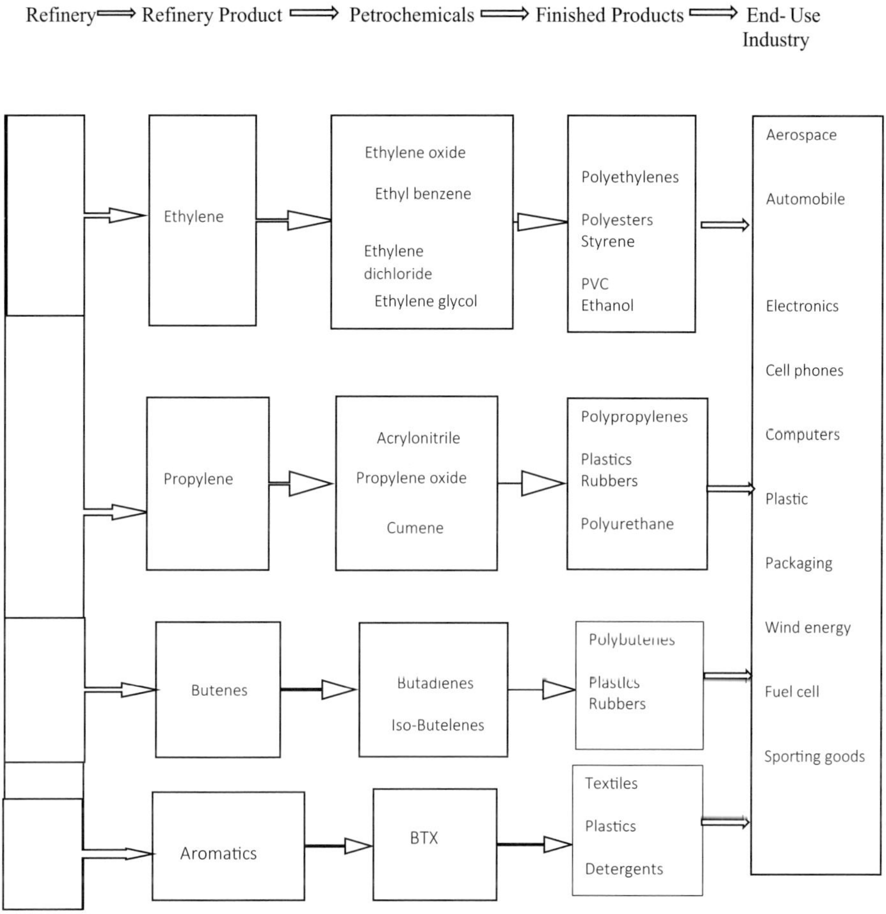

Fig. 6.3. A snapshot of the pathways from paraffinic feedstocks to the final petrochemical end products

6.4 Technologies for the production of petrochemicals

6.4.1 Background

Recently, the global petrochemical industry has grown and almost exploded due to the growing opportunities in other industries as well. Petrochemical industries have been in existence for more than a century, but recently, technological advancement has contributed dramatically in the world economy. According to several EIA reports, the chemical industry in the United States has grown to produce about half a trillion dollars' worth of products annually and to export about $100 billion of those materials. With growing worldwide populations and their increasing standard of living in all aspects of life, the demand for petroleum products is expected to increase further with no sign of slowing down. That's why this book addresses an important and timely issue for the petrochemical and energy industries.

The most challenging factor, however, is the choice of proper technology. Among the various technologies available to the manufacturer for the production of petrochemicals, it could be highlighted that the production of lower olefins through dehydrogenation of lower paraffins (C_2–C_4) may be the most suitable route for this application.

Dehydrogenation involves removal of hydrogen from saturated hydrocarbons, as follows:

$$\underset{\text{Paraffin}}{R_n\text{-}CH_2\text{-}CH_3} \quad \leftrightarrow \quad \underset{\text{Olefin}}{R_n\text{-}CH{=}CH2} \quad + \quad H2$$

where R is an alkyl group of chain length 'n' number of carbon atoms. Thus, the process is not only producing olefins but also hydrogen as a byproduct. This hydrogen can be reutilized within the refinery by recycling back to a hydrotreater or hydrocracker.

Dehydrogenation, being endothermic in nature, is quite energy-intensive. Another route available for the industry with less energy consumption is called oxidative dehydrogenation, which occurs in the presence of oxygen, as follows:

$$R_n\text{-}CH_2\text{-}CH_3 + \tfrac{1}{2}O_2 \quad \leftrightarrow \quad R_n\text{-}CH{=}CH2 \quad + \quad H2O$$

This reaction being exothermic requires less energy than the traditional dehydrogenation process, but the main problem is the strong probability of occurrence of paraffin combustion against the dehydrogenation reaction. Hence, the reaction may become uncontrollable, with the probability of an expensive shutdown.

The main source for the lower paraffinic feedstocks for the process is crude oil or natural gas. Lower hydrocarbons (C_2–C_4) are separated physically in a fractionation column inside the refinery. Cost of separation of individual hydrocarbons increases with the complexity of the mixtures and the purity of the target hydrocarbons required.

Due to the economic advantages, the energy industries are turning their investments more towards petrochemical industries than any other fields. Hence, there is also increase in continuing research activities for the development of new technologies. According to a 2018 Petrochemical report, almost one third of the chemical production technologies are based on steam cracking, and another quarter on oxidation of hydrocarbons. They are the two dominating technologies, with dehydrogenation being far behind, occupying less than 10% of the total industrial technologies.

It is becoming increasingly challenging for researchers to develop suitable catalyst to carry out the process at a minimum cost. The dehydrogenation reactions include high temperatures (above 600 °C) and mild operating pressures (below 100 psi). The catalyst normally applied to reduce the reaction temperature in dehydrogenation reactions is based on metals impregnated with alumina.

The recent increase in the production of tight gas/shale gas and the resulting increased availability of a cheap ethane source means the whole North American scenario is changing rapidly. Now, ethane is the biggest source of feedstock for ethylene production, and all steam cracker plants are switching their feedstock to ethane from naphtha. Naphtha is now available more for gasoline blends. The feedstock cost advantage is much higher in the US, as the cost of naphtha is almost three times that of ethane in equivalent mass basis.

There are 36 steam crackers in the US, out of which 33 are in the Gulf region, whereas Canada has 6, out of which 4 are in Alberta[2]. Thus, the northeastern area of Marcellus and Utica are essentially without access to any cracker plant. There is a serious shortage of pipeline for transportation. They have only two options: (i) either build a pipeline to the Gulf Coast plants, or (ii) build cracker plants in the nearby New York or Pennsylvania area. However, economics only dictates the solution, not the politics.

It should be cautioned here that the evaluation of new technologies is not an easy task. The most important factor in assessing a new technology is the "cost" of production. As this author has mentioned previously in his earlier book on heavy oil, "Economics always beats technology." Next to the cost of production comes the market value of the product. It not only depends on the market demand but also the geographical location, i.e., transportation cost. There are discussions going on for new full-scale crackers in the Eastern zone. But that needs a large-scale investment. The advantage is that it is closer to the feedstock, and it is cheaper to transport solid polymer finished product than gaseous feedstock.

There is an increasing demand for ethylene for polyethylene production. The obvious choice should be "dehydrogenation of ethane," as simple as it sounds. But even today, there is no attractive technology available that solves all the problems associated with the process. Rather, dehydrogenation of propane to propylene is much easier from the technical point of view. As has been discussed previously in Chapters 3 and 4, ethane dehydrogenation requires a higher temperature than with propane dehydrogenation. Commercial steam cracker furnace temps decrease with the increase in molecular mass of the feedstocks—for example, the "Coil outlet temperature - COT" for C_2 hydrocarbon is between 840°–850 °C and for C_3, it is 830–840 °C, and for C_4 and higher hydrocarbons, COT is <820 °C at a steam to oil ratio of 0.35.

When the severity of the process increases, though the conversion increases, the selectivity towards ethylene formation decreases (see Figure 4.6.1), with the increase in secondary products and including fouling due to coking. If the feedstock gets heavier or for mixed feedstock, the problems with separation and the additional cost for purity of the final product also multiplies. That is the reason, though naphtha cracking sounds attractive in some parts of the world (not in the US), that due to the poor selectivity towards ethylene, high conversion naphtha cracking is not an economical solution.

Though steam cracking is popular, other technologies are also available for the production of olefins from paraffins, such as a partial-oxidation process, as discussed above. This sounds attractive over the dehydrogenation process, which is endothermic and requires heat; oxidation is exothermic and driven by the heat produced from the system itself. But this process also has its own negative points: (i) most importantly, it produces CO_2, (ii) selectivity towards olefin is lower than that of the steam cracking process, and (iii) finally, the oxidation process is difficult to control and can overshoot towards the formation of coke at any time, with shutdown of the whole plant at a very high economical penalty.

Propylene, the most popular petrochemical today, is primarily manufactured in the refinery by steam cracking or catalytic cracking of C_3 or higher molecular weight feedstocks. A steam cracker produces "polymer grade" propylene, the quality depending on the quality of the feedstock.

Petrochemicals are produced from naphtha, which is obtained after fractionation of crude oil, and then steam cracking. It would have been more favorable if they were produced directly from the crude oil itself with minimum treatment. For example, according to the author's internal studies, tight crude oil from the major US sources contains aromatic petrochemicals, such as benzene, ethylbenzene, and xylenes (no toluene). Especially the tight oil from the Permian basin consists of higher concentrations of aromatics, between 3 to 5 volume percentages of the total crude. The amounts of individual components depends on the crude oil location.

6.4.2 Crude to chemicals

It is not easy to identify the best technology for any particular process. Every project has its own unique risk, not only because of its complexities but also due to the geographical location. Risk factors also include the local environmental regulations. The best result would be to do comparative studies of various processes and to identify the impact locally (or where the product is sold).

As the demand for petrochemicals is expected to increase beyond 2020, the competition is increasing with the increase in the number of players in this industry. Several reports indicate a wave of steam cracker plants is coming up in the near future around the world, in particular, several ethane-cracker plants in the United States, to produce polyethylenes plastics of the type LDPE/LLDPE, mostly for export. Refineries are also looking for alternate technologies to remain competitive. In order to increase their profitability, one technology that is getting a lot of interest is direct conversion of crude to petrochemicals.

Some refinery technologies such as fluid catalytic cracking (FCC), delayed coking, or catalytic reforming allow the refiners the flexibility to operate to increase the production of olefins for the petrochemicals industry. Nevertheless, this requires higher severity and hence higher operating cost. Despite the higher cost, the alternate option may be profitable as the demand for petrochemicals increases. Some of the refinery technologies are discussed below.

6.4.2.1 HS-FCC process by Axens

In some cases, of course, based on economics, some refineries produce propylene as a byproduct of an FCC plant. The main purpose of an FCC plant in a refinery is to produce gasoline from heavy gas oil feedstock. So it is a challenge for the refinery to tweak the operating conditions in favor of propylene instead of gasoline, and then redirect the output line to a "propylene splitter" to produce polymer-grade propylene for the petrochemical industry. However, in the US, with increasing demand for gasoline and an increase in cheap lower hydrocarbons with the production of tight gas, C_2 / C_3 steam crackers are in greater use for the production of olefins. In some parts of the world where gasoline demand is high, they cannot afford to switch the production from gasoline to olefins. In a typical FCC plant, gaseous hydrocarbon (< C4) production could be 15 to 20 wt% of the feed, and the ethylene plus propylene production could exceed 30% of the gaseous components.

According to the Axens company website (look for the English version), due to the increasing demand for petrochemicals in Europe and the corresponding lesser availability of feedstocks at reasonable cost, they have improved the existing FCC process to maximize olefin production from the same FCC feed VGO (vacuum gas oil). The process is known as "High Severity FCC or HS-FCC ™," where they have increased the severity of the reaction conditions. Some of the selective results taken from the Axens website are shown in Table 6.4.2.1. The results clearly agree with the discussion in earlier chapters, where it says that the dehydrogenation of paraffin for olefin production increases at higher temperatures though at a higher cost. Figure 6.4.2.1 represents a simplified version of the process, redrawn from the flow diagram shown on their website.

Table 6.4.2.1 shows typical average results generated by Axens in their HS-FCC plant and comparison data with a conventional FCC plant. In the HS-FCC process, by simply increasing the reactor temperature by 100 °C, the ethylene production increased from 1.5% to 2.5% and the propylene production tripled from 5% to 16%, at the cost of a decrease in gasoline production by about 15%. By improving the catalyst performance with the increase in VGO conversion, Axens claims to achieve almost 20% propylene yield. Axens not only concentrates on the production of polymer-grade ethylene and propylene, but also on the aromatic petrochemicals as a byproduct.

Table 6.4.2.1. Average yield (wt%) of major olefins from HS-FCC as compared to the conventional FCC

	FCC (conventional)	HS-FCC
Temp, °C	500	600
Ethylene	1.5%	2.5%
Propylene	5.0%	16.0%
Gasoline	51.5%	38.0%
Coke	4.5%	6.5%

Figure 6.4.2.1 shows that they have an additional ethane steam cracker in the front of the HS-FCC plant for additional production of ethylene. This clearly suggests that there is so much demand for ethylene in Europe that it is considered as a high-performance product such that they can afford to run integrated HS-FCC and steam cracker plants.

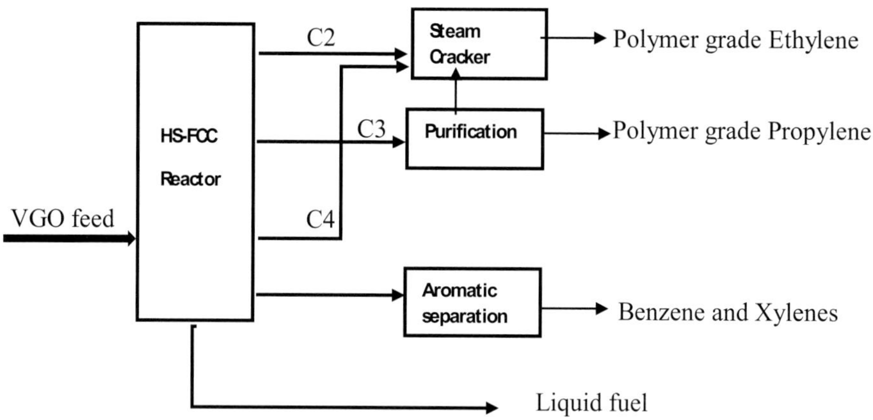

Fig. 6.4.2.1. Redrawn simplified schematic diagram of Axens HS-FCC plant

6.4.2.2 Petro-FCC™ process by UOP

In order to meet the world demand, UOP has also introduced a new option with the modification of their existing FCC process called Petro-FCC™ (Fig 6.4.2.2). In this case, the conventional FCC mode for the production of gasoline is modified to maximize the propylene yield with the introduction of several improved steps: catalyst type, reactor, regenerator, and olefin separation sections. According to the UOP website, propylene yield increases to about 20%. They also claim to have the capability to control the propylene yield based on market demand.

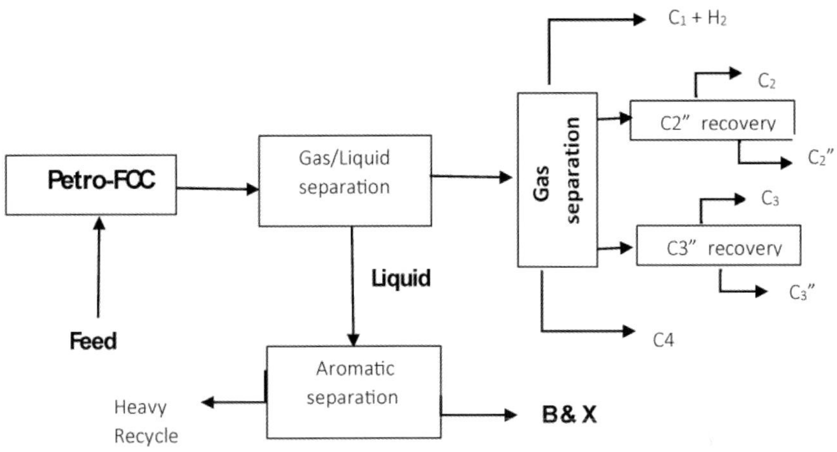

Fig. 6.4.2.2. Redrawn simplified schematic diagram of Petro-FCC plant

6.4.2.3 Delayed coker plant

Typically, a delayed coker is used in a refinery to treat difficult-to-handle heavier feedstocks, such as heavy oil, bitumen, atmospheric resid, or vacuum resid. The main purpose of the delayed coker is not the production of olefins, but it could be a by-product for the petrochemical industries, as discussed below. The details of the delayed coking process are discussed in the next chapter.

Similar to the FCC plant discussed above, in a refinery, a delayed coker plant could also directly produce olefins from the heavy crude oil feedstocks because it also runs at a very high temperature (> 900 °F / > 480 °C). It is to be noted here that the FCC runs with catalyst, and the delayed coker is a thermal process where no catalyst is involved. We already know that higher temperatures promote olefin formation.

Figure 6.4.2.3 illustrates the results of a delayed coking process carried out at 930 °F (500 °C) using a typical vacuum resid feedstock (source: author). In this process, an average C_1–C_4 gaseous hydrocarbons yield was 9.5 wt% of the feed. The figure represents the molar ratios of C_2 to C_4 hydrocarbons (including paraffins and olefins) with respect to methane in order to eliminate the effect of operational or feedstock quality variables. Methane is always present at the highest concentration in a delayed coker off-gas stream, and it is almost half of the total moles of gaseous hydrocarbons (C_1–C_4) produced. Results show that the yields of gaseous hydrocarbons decrease with the increase in molecular weight of the hydrocarbons from C_2 to C_4.

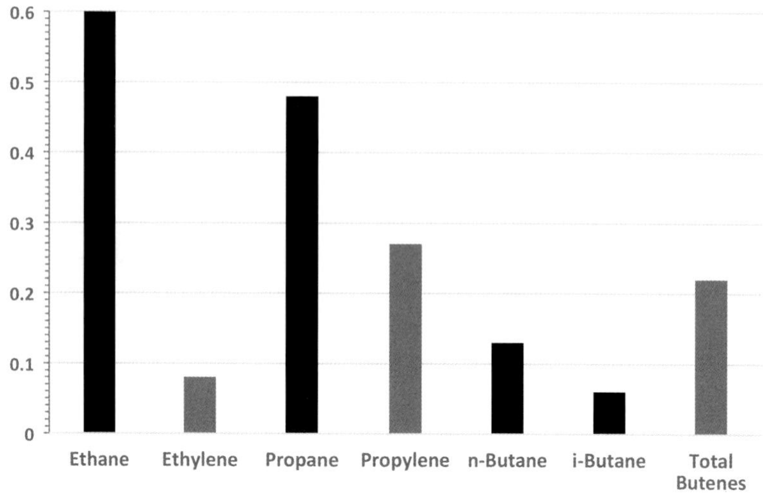

Fig. 6.4.2.3. Off-gas composition with respect to methane from delayed coking of a typical vacuum resid at 930°F

Figure 6.4.2.4 represents the molar ratios of olefins to paraffins of individual C_2 to C_4 hydrocarbons in the gaseous stream. It is clear that the amount of olefin production with respect to its corresponding paraffin increases with the increase in molecular weight of the hydrocarbons. In the case of C_2 hydrocarbons, ethylene yield is almost $1/10^{th}$ of the ethane and for C_3 hydrocarbons, propylene yield is almost half of the propane. On the other hand, in the case of C_4 hydrocarbons, total butenes (butene-1, butene-2, iso-butene, and butylenes) production is almost 10% higher than that of total butanes (normal + iso) formation. The results clearly concur with the chemical reaction mechanisms discussed in Chapters 1 and 2, which explains why it is easier to dehydrogenate bigger molecules as compared to the smaller molecules under the same temperature and reaction conditions.

In Table 6.4.2.1, Axens data show that in a conventional FCC plant (using VGO feed) at 500 °C (930 F), the production of propylene is almost 3.5 times that of ethylene and, surprisingly enough, in a delayed coker process almost at the same temperature with no catalyst and with heavier feedstock (vacuum resid), the propylene production with respect to that of ethylene is almost the same (Figure 6.4.2.4).

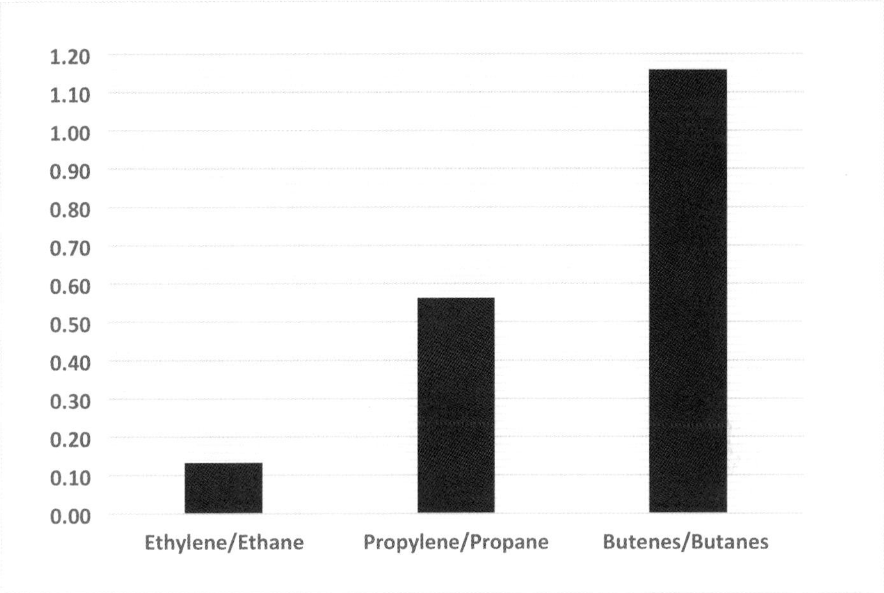

Fig. 6.4.2.4. Molar ratios of olefin to paraffins in off-gas from delayed coking of a typical vacuum resid at 930°F

6.4.2.4 Aromatics to petrochemicals

One of the most valuable sources for the petrochemical industry after olefins is aromatics. Among the aromatics, benzene and para-xylene are of the highest value. Para-xylene is used as feedstock for the manufacture of terephthalic acid, which in turn becomes the raw material for the manufacture of plastics. Reliance, in India, is one of the biggest manufacturers of plastics in the world via this route.

These aromatics are available in relatively smaller quantities in the virgin naphtha-range crude oil. However, with growing demand for aromatics, refiners are now conducting catalytic reforming processes in conjunction with other technologies to produce additional lighter aromatics that are of high added value to improve their economics. Two well-known catalytic reforming technologies are: (i) the CCR Platforming™ process by UOP and (ii) the Aromizing™ process by Axens.

Usually, refiners direct the reformed naphtha containing the light aromatics to the gasoline pool, without aromatics separation. However, due to the high demand for aromatics as intermediates, petrochemical refiners now install aromatics separation plants, as presented in Figure 6.4.2.5.

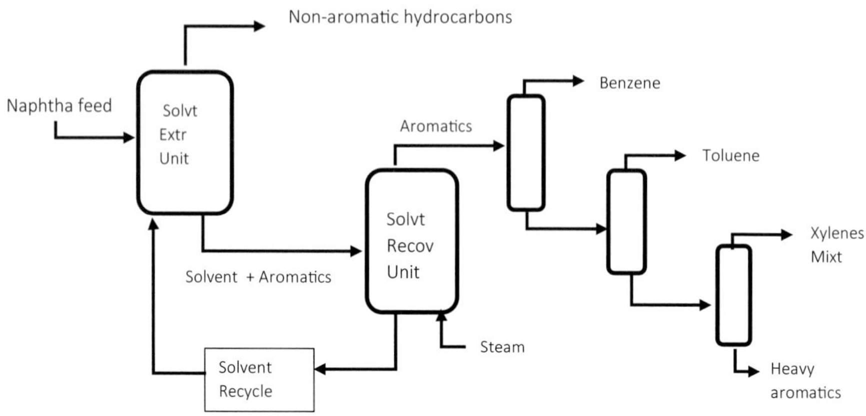

Fig. 6.4.2.5. Redrawn simplified schematic diagram of a typical aromatics separation plant

The process is based on a solvent extraction technique[3], whereby the aromatics are preferentially and selectively separated from the naphtha fraction. In the first vessel, the feed and the solvent are introduced and well mixed. Aromatics get dissolved into the solvent and recovered from the bottom of the vessel, and the non-aromatic hydrocarbon mixture is recovered from the top.

The solvent rich in aromatics is then fed to a second vessel where the solvent is separated from the aromatics. The recovered solvent is recycled back. The enriched aromatic stream is then fed to three successive aromatic (BTX) recovery columns to separate benzene, toluene, and xylenes from other heavier aromatics in the mixture. The individual aromatics are then shipped to petrochemical industries for further treatment.

6.4.3 Steam cracker plant

As mentioned earlier, the steam cracker is the most popular technology for the conversion of lower hydrocarbons (C_2–C_5) into olefins as petrochemical feedstocks. Though widely used, this is also the most energy-intensive technology. The conversion of lighter hydrocarbons to higher-value olefins proceeds in the vapor phase as discussed in earlier chapters. The rate of reactions varies with the molecular size and type of the reactant, as follows:

$$\text{n-paraffins} > \text{iso-paraffins} > \text{cyclo-paraffins}$$

and among the paraffins, the reactivity increases with the increase in carbon number in the molecules.

Of course, depending on product demand and economic interest, individual hydrocarbons will be treated differently, maybe with different technologies. For example, in the US, utilizing methane as fuel inside the refinery may be more economical than converting it into petrochemicals. Due to the higher concentration of ethane in the shale gaseous stream, there is an abundance of ethane in the US; hence, it is more attractive as steam cracker feed to produce ethylene than the liquid hydrocarbons. Similarly, propane is more attractive to produce propylene and butane for C_4 olefins because of increase in selectivity.

Each individual paraffin is fed to the steam cracker, mixed with high-temperature steam, and then further heated at a higher temperature (> 600 °C) inside the furnace. As the product stream exits the furnace, it is cooled rapidly in steps and then sent to another fractionator to separate hydrocarbons from the target olefins. The final olefin is then transported to the market for further treatment.

The whole process of olefin production with a steam cracker consists of three major steps:

(i) Thermal reactor/pyrolysis plant—this is the heart of the process, where super-heated steam and hydrocarbon feedstock enter a tubular reactor. Usually, a tubular reactor is somewhere between 20 to 25 meters long (10 to 15 cm id) and heated between 700 to 900 °C. Such a high temperature of the furnace is achieved by placing the tubular reactor inside a gas-fired burner. Pyrolysis of the hydrocarbon takes place depending on the severity required based on the type of feedstock. As explained in previous chapters, at higher temperatures, thermal cracking followed by a dehydrogenation reaction takes place through a free radical reaction mechanism, converting paraffinic hydrocarbons into olefinic hydrocarbons.

(ii) The second step is at the exit of the tubular reactor, where the hot product is cooled in several steps with a series of heat exchangers bringing the temperature down to somewhere around 300 to 350 °C. The cooling steps serve two purposes: (a) avoids unwanted secondary reactions of the olefins and (b) produces super-heated steam as a byproduct through heat exchangers.

(iii) The third important step is the separation and purification of the desired olefin product. The higher the molecular weight of the feedstock or mixture of hydrocarbons, such as in naphtha, the more difficult and expensive is the separation step.

Since the beginning of the first steam cracker developed by Kellog[4] in the early 1960s, the application of this technology has increased and, especially recently, exploded. The original first-generation steam cracker used liquid naphtha hydrocarbon as feed that was preheated to 650-690 °C, and steam was preheated to 930-950 °C separately. Those two were fed to an adiabatic reactor set at a pressure of

2½ atmospheres. The outlet of the reactor was cooled to transfer steam into water and separated first. Then the hydrocarbon product stream was fed to a fractionator to separate the desired products—ethylene or propylene—from the paraffins and unreacted heavier hydrocarbons. Kellog reported that in a commercial operation, the average yield of ethylene from a heavy naphtha fraction varied between 30 to 35 wt% of the feed.

Figures 6.4.3.1 and 6.4.3.2 show the results of the production of C_2 to C_4 olefins from the steam cracking of ethane, propane, butane, and naphtha feedstocks.

Due to the higher production of ethylene as compared to the other olefins (< 15%), they were separated into two figures for clear illustration within the scale.

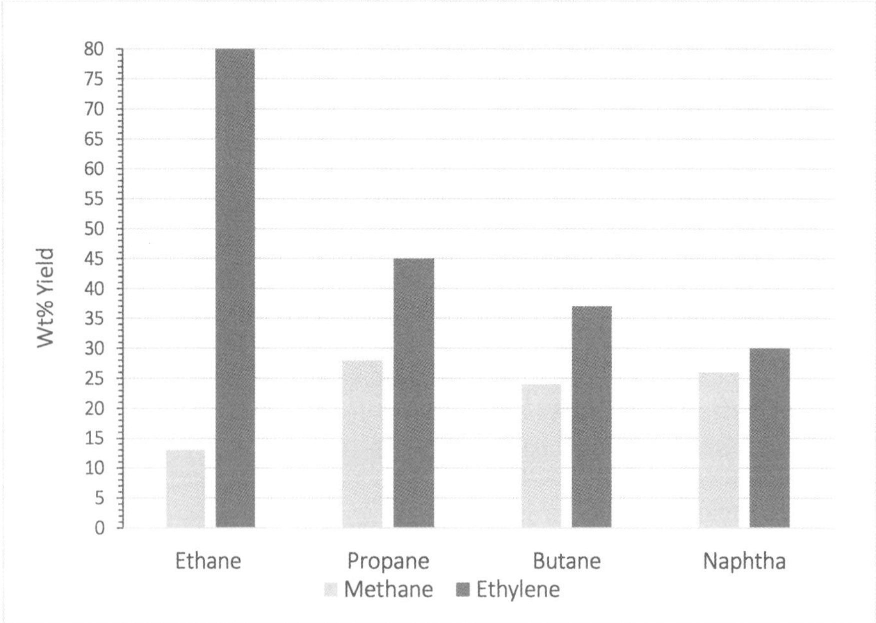

Fig. 6.4.3.1. Methane and ethylene yields from the steam cracking of various paraffinic hydrocarbon feedstocks

As shown in Figure 6.4.3.1, the ethylene yield is 80% from ethane as feedstock and decreases with the increase in molecular weight of the feedstock. Naphtha, which is a mixture of hydrocarbons, produces only 30% ethylene, lower than the other paraffins. However, it is interesting to see that the methane yield remains almost the same, closer to 25 wt% for all three feedstocks, except for ethane where it decreased to 10%. This is expected; as we discussed earlier, in a free radical reaction mechanism, the cracking reaction is dependent on the bond energies, and it is more difficult to break the C-C bond of ethane to produce methane than the higher hydrocarbons.

Figure 6.4.3.2 illustrates the production of propylene, butadiene, and butenes (mixture) from ethane, propane, butane, and naphtha feedstocks, respectively. The reaction mechanism clearly indicates that the production of olefins of higher molecular weight than the feedstock itself is not favorable under the steam cracking process. The results show that the production of propylene is mostly favored from higher molecular weight (> C_3) feed, and the yield is almost the same (13+/- 1%) as that from the propane itself. Dehydrogenation and cracking are two competing reactions occurring at the same time in the case of C_4 and naphtha feedstocks, as butane yields more C_3-olefins than C_4-olefins.

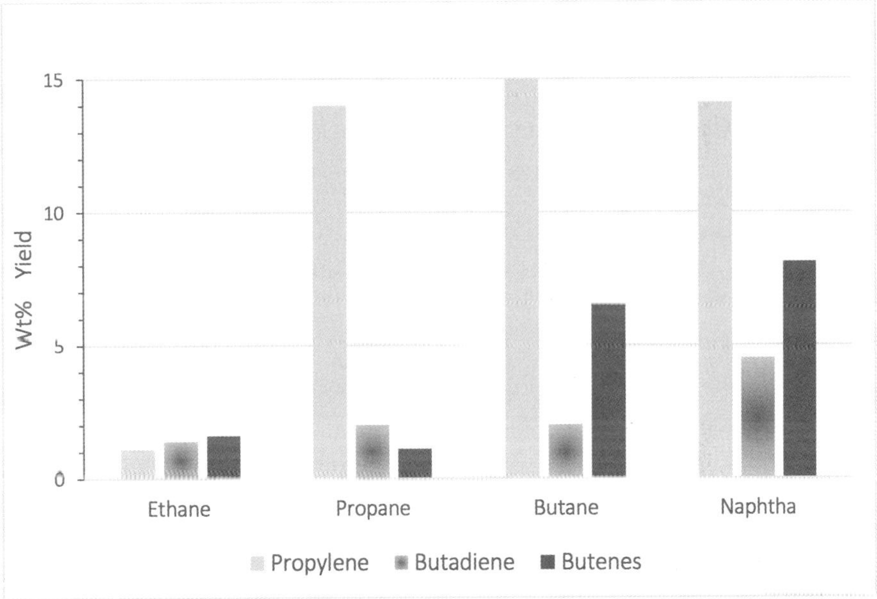

Fig. 6.4.3.2. C_3 and C_4 olefin yields from steam cracking of various paraffinic hydrocarbon feedstocks

Further, results also show that the butane produces almost three times more butenes than butadiene, whereas a higher molecular weight naphtha feed favors double dehydrogenation to butadiene, with almost double (5%) yield than the butane feed. These results support the free radical reaction mechanisms discussed in earlier chapters.

Commercial value of the olefins for the petrochemical industries is more than double the market value of the corresponding paraffins.

6.5 Commercial technologies by various companies

There are several companies in the petrochemical industry claiming proprietary technology for producing olefins, basically utilizing steam cracker with or without catalysts. Due to confidentiality reasons, we do not know the details of the process design and their significant differences. Some of the technologies are discussed below, utilizing major sources of publicly available information, such as Oil & Gas Journal, Hydrocarbon Processing, individual company websites, or other technical websites. Snapshots of the major processes from various sources are summarized here; however, readers interested in detail may have to contact the respective technology developer or licensors. Readers should also note that in the present case, the author selected a few commercial processes, but the author is neither recommending nor promoting any of the technologies discussed below. According to the author's view, the economics of any process is highly site-specific, and one should not depend on the economic values provided by the company or any other institution, especially for advertisement purposes. One should calculate detailed economics for themselves for a particular project, at a particular location, for a specific technology, and specific feedstock.

6.5.1 Paraffin dehydrogenation (PDH) process

Although earlier chapters discussed in detail thermal processes without catalyst, and especially dehydrogenation of paraffins in a thermal tubular reactor, industry usually prefers catalytic processes that produce higher amounts of desired olefins with high selectivity. This comes, though, at a cost of the catalyst and frequent regeneration of the catalyst due to the very high rate of deactivation. Various companies have developed a process of their own with different catalysts, regeneration methods, and purification of the final product.

Figure 6.5.1 represents a general snapshot of the basic principles of the PDH process, with a simplified process flow diagram for a single pass conversion. Fresh paraffin feed is mixed with recycled paraffin from the olefin-paraffin splitter. This paraffin stream is first send to a de-oiler to remove impurities (not shown in figure), before feeding to the reactor. The feed is added from the top of the catalytic reactor.

Steam is added to the hydrocarbon feed to reduce the partial pressure of the reactant in order to increase the dehydrogenation reaction at a given temperature. Steam also reduces coke formation on the catalyst bed, thus decreasing the deactivation rate of the catalyst and hence increasing the regeneration cycle time. The process consists of a fixed bed catalyst reactor that is externally heated with a natural gas furnace that provides the heat required for the endothermic reactions.

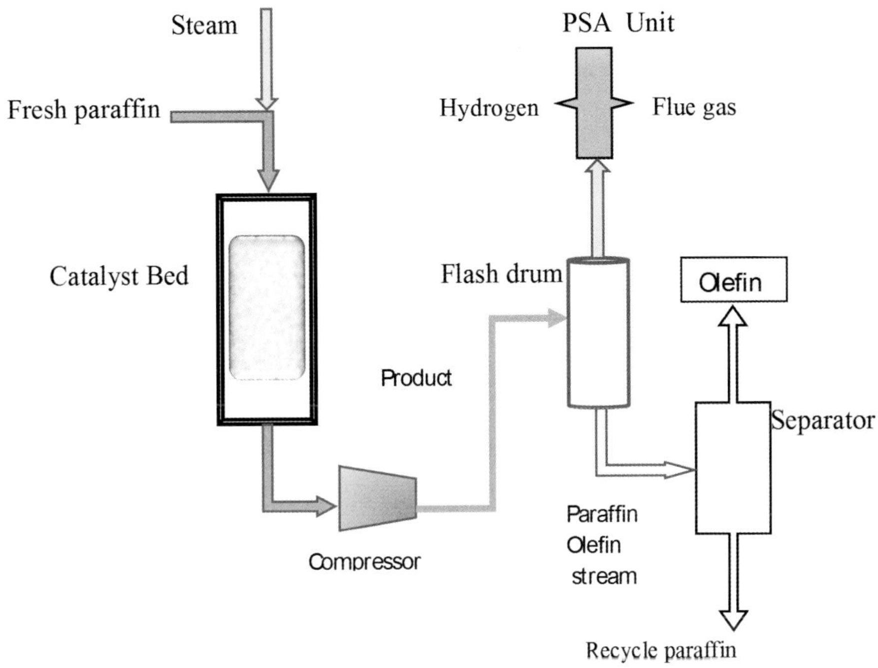

Fig. 6.5.1. Basic flow diagram of a general PDH plant

In general, all PDH reactors run between 550 °C and 650 °C at atmospheric or below atmospheric pressure. The reactor temperature is controlled in order to conduct the reactions at isothermal conditions to avoid unwanted secondary reactions. Most of the companies use this step as proprietary information. Though the figure shows only one reactor, every company uses multiple reactors of different sizes, depending on their own economics that dictates the design of the equipment.

The product stream is then compressed and directed to a flash drum where hydrogen and lighter flue gases are separated. The hydrogen is separated and purified in a PSA (pressure swing adsorption) unit. The unreacted paraffin and product olefin mixture is fed to a fractionator, where those two components are separated. Paraffins are recycled back and the product olefin is sold to a petrochemical industry for further treatment with a polymerization step to attain the desired plastic products.

The flash drum works at a very low negative temperature at a certain pressure to maintain the vapor–liquid equilibrium, the condition depending on the type of hydrocarbon. For example, for ethylene, the atmospheric equivalent boiling point temperature is -104 °C. Flashing is done simply by letting down the pressure.

Petal Consulting Company published a report in the O&G J (Sept 2018) showing ethylene production costs in the US. According to the study, olefin production cost is determined by the feedstock cost and the credit from the byproducts. The byproduct yield varies significantly with the types of feedstocks: pure ethane, mixed gaseous feed, or liquid feed. Raw material cost is determined from the cost of the paraffin feedstock and its conversion to ethylene. Byproduct credit is calculated by the spot market price of those materials, depending on its purity. According to the report, 80% of the ethylene production in the US in 2018 was from ethane, and the cost varied between 12 to 13 cents per pound. The production cost in the US is considerably lower than that of Europe and Asia.

Though there are several paraffin dehydrogenation processes, developed by various companies, two major technologies Catofin and Oleflex are used commercially in North America, as described below. STAR technology, which is actually an extension of the PDH process, is also a commercial process running in the US.

6.5.1.1 PDH process—licensed by BASF

BASF, in collaboration with Stat Oil of Norway, commercialized this technology with the name of PDH process. The technology is similar to the general overview given above. It is based on a fixed-bed reactor packed with chromium oxide/alumina catalyst, run at a temperature between 590 °C to 600 °C, and at atmospheric pressure.

6.5.1.2 CATOFIN™—developed by Lummus and licensed by CB&I

Lummus has developed a patented commercial process called Catofin, based on catalytic dehydrogenation of propane to propylene and iso-butane to produce is-butylene. Publicly available information shows that the process utilizes a fixed-bed catalytic reactor, with chromium supported on alumina (Cr_2O_3 / Al_2O_3) as catalyst. The dehydrogenation reaction, being endothermic, requires higher temperature and equilibrium limited. On the other hand, higher temperature is not favorable for the process as selectivity of the required olefins decreases and the most unwanted reaction—coking—accelerates the deactivation rate of the catalyst. The operation becomes difficult to control with the increase in catalyst regeneration steps within a short period of time. In order to make the process continuous, the CATOFIN process uses several reactors operating in cycles; while one reactor is down for catalyst regeneration, another reactor is in use. Reactors are operated around 650°C, at close to atmospheric pressure, achieving 50 to 70% conversion, with very high selectivity of close to 90%.

6.5.1.3 OLEFLEX™—UOP Technology

UOP has also developed a similar dehydrogenation process named OLEFLEX. The difference from the Lummus process is that they claim to use platinum and tin as the dehydrogenation catalysts, supported on a unique spherical alumina base. The activity of platinum is favored at higher temperatures as compared to any other metals, thus reducing the time in the regeneration step, though at a certain cost of the metal itself. This technology is a combination of two earlier technologies developed by UOP:

(i) Pacol process for the conversion of paraffins to mono-olefins, and
(ii) Platforming based on platinum catalyst

The process uses three reactors in series; though it is considered to be a fixed-bed process, in this scenario, UOP actually moves the catalyst bed from one reactor to another, and basically makes it a moving-bed system. The whole system is capable of producing olefins continuously with a continuous catalyst regeneration step. The feed is preheated before entering each reactor. They use sufficient hydrogen to reduce catalyst poisoning and to help reduce the frequency of regeneration and hence the cost of operation.

In a typical commercial plant with capacity of around 650,000 tons/year, polymer-grade propylene is produced from pure propane with overall propylene yield of 90 wt% at a temperature around 650 °C and at near atmospheric pressure.

6.5.2 Oxidative dehydrogenation process

The oxidative dehydrogenation process was developed to produce olefins at a higher energy efficiency as compared to the PDH process. Though the basic principle is the same as that of the PDH described above, there is a difference in process design in order to accelerate the overall dehydrogenation step. As it is known that the dehydrogenation of paraffin is equilibrium dependent and it reaches a maximum at a certain temperature and pressure, the equilibrium is shifted with the addition of an oxidative reactor in front of the dehydrogenation reactor. The shift in the equilibrium favors additional dehydrogenation. The feedstock is always gaseous hydrocarbons in order to increase olefin production. Because oxidation is exothermic and produces heat, it saves on energy required to do the same step to achieve the same conversion. But the process requires highly pure oxygen (>90%), and hence the cost of oxygen increases the operating cost.

6.5.2.1 STAR™—commercialized by ThyssenKrupp-Uhde of Germany

The STAR (Steam Active Reforming) process was the first to apply an additional reactor in front of the STAR PDH reactor, called an oxy-dehydrogenation reactor. It is based on the fact that it shifts the equilibrium thermodynamic of propane dehydrogenation in favor of propylene formation with the addition of oxygen. In the first step STAR PDH reactor, olefins are produced with the formation of hydrogen that limits the reaction equilibrium. In the second reactor, a small amount of oxygen is added to convert the hydrogen to water, thus shifting the equilibrium towards the formation of olefins.

Figure 6.5.2 illustrates a simplified schematic diagram of the STAR process, redrawn from their company website. According to their information, the commercial operation is designed for the manufacture of propylene in a Propylene-Oxidative-Dehydrogenation process. They use metals impregnated on zinc-calcium alumina support in a fixed-bed reactor. The STAR process uses heat-resistant metals that can tolerate the reaction conditions as the dehydrogenation catalyst. The process utilizes pure propane as feedstock, and the reactor temperature is between 550 °C - 600 °C.

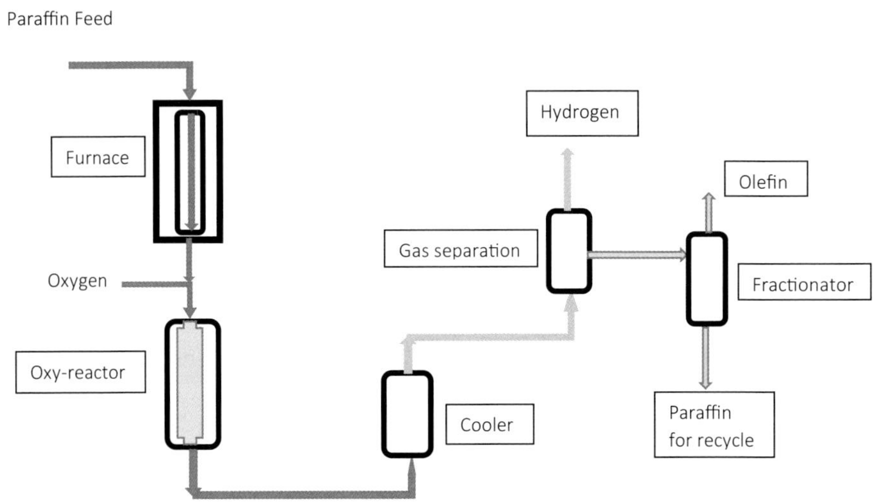

Fig. 6.5.2. Simplified schematic diagram of Uhde STAR technology (redrawn from the company website figure)

The active element of the catalyst system is platinum, which can withstand higher temperatures and has longer life before regeneration takes place. The STAR process also has application in the production of aromatics through dehydrocyclization of C7–C8 paraffins.

Table 6.5.2 provides a very good summary comparison of the three processes described above. This table was obtained from the web site of Digital Refining and was originally published in the journal Petroleum Technology Quarterly (2008).

Table 6.5.2. Comparison of three PDH processes

PDH process technology parameters	UOP Oleflex	Uhde STAR process	Lummus CATOFIN
Catalyst	Pt-Sn on alumina	Pt-Sn on zinc/calcium aluminate	Chromia on alumina
Reaction system	Fluidised bed system consisting of several radial-flow adiabatic reactors in series and an external regenerator (CCR)	Fixed-bed system consisting of an externally fired tubular reactor and an adiabatic reactor in series	Fixed-bed system consisting of a number of axial-flow adiabatic reactors connected in series
Main source for heat supply	Preheating of the feed and intermediate products before each reactor	External firing of the tubular reactor and selective hydrogen combustion in the adiabatic reactor	Heating the catalyst during catalyst regeneration
Operating mode	Continuous (5–10 days' catalyst cycle time)	Discontinuous (7 hours' normal operation, 1-hour regeneration)	Discontinuous (6–20 minutes' reaction cycle)
Pressure, bara	2	5	0.3–0.5
Temperature, °C	580–650	570–590	560–650
LHSV, [m_ liq. feed/(m_ catalyst * h)]	4	6	< 1
Per-pass conversion propane, %	35	40	48–65
Selectivity propene, mol-%	89–91	89	88

6.5.3 FBD™ process—marketed by SnamProgetti / Yarsintez

Though this is a dehydrogenation process, the main difference with the other technologies described above is that it uses a fluidized-bed reactor (FBD), unlike the other processes where a fixed-bed reactor is used. The process was originally developed in Russia (former Soviet Union) and is currently marketed by SnamProgetti partnered with Yarsintez of Italy.

The FBD process is very similar to the FCC unit (fluid catalytic cracking) used by almost every refinery to convert gas oil into gasoline and diesel. The only difference is that of the catalyst type used, where the catalyst is chromium oxide on alumina particles. The fluidized reactor runs at a temperature between 550-600 °C, at atmospheric pressure with super-heated steam. In this process, the spent catalyst is withdrawn continuously from the reactor, regenerated, and recycled back. In a regular FCC plant, the propylene yield from the gas oil feed is around 5 to 7%; however, by improving the catalyst and operating conditions, the yield could be increased by almost five fold.

SnamProgetti markets the FBD process under various versions of the technology depending on the desired product, as for example, the FBD-3 process for the production of propylene and FBD-4 for the production of iso-butylene. The FBD- 3 process is attractive for the refiners who want to use heavier feedstock, and FBD-4 is attractive for the manufacture of MTBE.

6.5.4 Methanol-to-Olefin (MTO) technology

With the increase in demand for olefin worldwide, there is also demand for alternate technologies besides steam crackers, utilizing lower-value feedstocks for the production of olefins. Especially in Europe and China, Methanol-to-Olefin

(MTO) technology is getting a lot of attention where natural gas is quite expensive. Methanol could be produced from lower-value carbon sources like coal or heavy crude oil via the syngas route. Coal or coke is converted into syngas through gasification technology (discussed below in subsection 6.5.8). Syngas ($CO + H_2$) could be converted into methanol through commercially known processes, as follows:

$$CO + 2\,H_2 \longrightarrow CH_3OH$$

The major steps involved in the MTO process to produce polyolefins are shown below:

Heavy hydrocarbon / Coal → Syngas CO + H2 → Alcohol

Alcohol → Olefin → Polyolefin

For example:

Methanol → Ethylene → Butene-1, Hexene-1, Polyethylene

Methanol is converted into various olefins or a mixture of olefins according to the following reactions:

(i) $2\,CH_3OH \longrightarrow C_2H_4 + 2H_2O$ (Ethylene formation)

(ii) $3\,CH_3OH \longrightarrow C_3H_6 + 3H_2O$ (Propylene formation)

(iii) $4\,CH_3OH \longrightarrow C_4H_8 + 4H_2O$ (mixture of butenes)

These reactions are very temperature-sensitive, and control of the reaction temperature is very important in order to achieve maximum selectivity of the desired olefin. For example, if the temperature is lower, higher molecular weight butenes will form as compared to the lower molecular weight propylene or ethylene.

Conversion of methanol is favored on an acid-based catalyst. In the presence of silica alumina (Zeolite)-based SAPO34 commercial catalyst, selectivity towards ethylene is higher than that of propylene, and methanol conversion increases from 70% at 400 °C to over 98% at 500 °C.

There are several companies competing to commercialize this technology; two examples of commercialized processes are given below:

6.5.4.1 MTO—Lummus process

Major olefins produced from the MTO process are ethylene and propylene along with some C_4 and higher olefins. Thus, the major challenge in this process is the separation of desired olefins, namely ethylene and propylene at a very high purity, from the mixture of hydrocarbons. Lummus claims that they developed proprietary technology for olefin separation from the mixture of hydrocarbons at the outlet of their MTO reactor. This light olefin recovery technology has the capability to recover better than 99% of the ethylene and propylene from the mixture, thus making the process efficient and cost-effective.

6.5.4.2 MTO—developed by UOP / Hydro

According to the publicly available information from UOP, they have developed a new generation of catalyst to make the MTO process highly efficient, with methanol conversion close to 100% and ethylene and propylene production close to 80% with selectivity towards the olefins (C2" + C3") better than 85%. They claimed to obtain better than 97% purity of those olefins, making it polymer-grade product. UOP's catalytic process claims to have other advantages, like controlling the ethylene to propylene ratio anywhere from 0.5 to 1.5, and thus controlling the types of olefin yields based on market demand. The process has the ability to crack the unwanted heavier (C5+") olefins to lower olefins (C2" + C3") utilizing an additional integrated olefin cracker unit.

6.5.5 Propylene production via metathesis

With increasing worldwide demand for propylene (even more than that for ethylene), the industry started looking for different approaches to meet the demand. The newer option called metathesis looked promising, as a refinery could use ethylene and butenes either from the steam cracker plant or from the FCC plant, or from both, to convert into propylene. Because this process could be integrated easily with the existing source of other olefins, it is cost-effective and efficient. The process can maximize the production of propylene to close to 90%, and hence may not require an expensive separation process, and is capable of producing polymer-grade propylene.

In this case, the carbon = carbon double bonds are broken and then rearranged to form propylene. Nevertheless, this technology could be used with varieties of C3, C4, or C5 olefins.

The basic reaction mechanism of metathesis is as follows:

$$\underset{\text{Ethylene}}{C=C} + \underset{\text{Butene-2}}{C\text{-}C=C\text{-}C} \xrightarrow{\text{Catalyst}} \underset{\text{Propylene}}{2\,C\text{-}C=C}$$

A simplified schematic diagram of the metathesis process by ABB Lummus is shown in Figure 6.5.5, consisting mainly of two important sections:

(i) Metathesis reactor, and

(ii) The purification or separation section

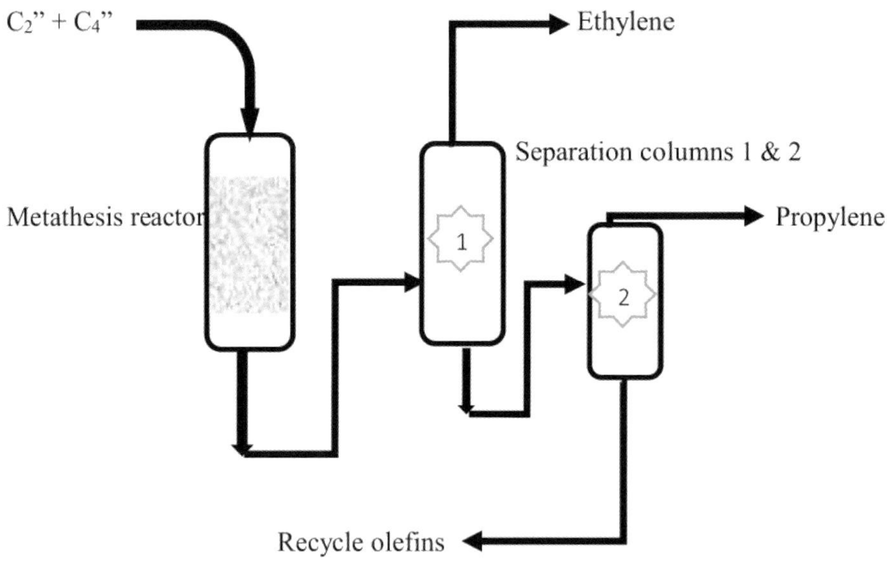

Fig. 6.5.5. Simplified schematic diagram of metathesis process

This process is known as Olefin Conversion Technology (OCT). According to the Lummus website, the feedstock consists of a mixture of ethylene and butenes (also recycled olefins), first cleaning from the material the potential for catalyst poisoning such as oxygenates and sulfur compounds (not shown). After the pretreatment, the stream is vaporized and heated to the reactor temperature and fed to the metathesis reactor. The product effluent is sent first to the ethylene recovery unit separator #1. The bottom stream, consisting of C3 + hydrocarbons, is sent to separator# 2, where propylene is recovered, and other hydrocarbons are recycled back.

Originally, this process was developed by Phillips Oil Co, where the reactor was packed with tungsten-based catalyst impregnated with magnesium oxide as co-catalyst on silica support. The process could achieve propylene production of greater than 90%.

6.5.6 Petcoke to petrochemicals

The alternate approach to produce petrochemicals from any low-value carbon-bearing feedstock is to convert the complex molecules into their most fundamental, simplest, energy-rich molecules carbon monoxide (CO) and hydrogen, known as syngas. Any solid carbon-rich source such as petcoke or even biomass could be converted into syngas through the process of gasification.

Gasification is a process of partial oxidation of carbon-rich hydrocarbon, carried out under controlled oxygen, at elevated temperatures (800 °C to 1500 °C) and pressures, according to the following chemical reaction:

$$CH_y + \frac{1}{2} O_2 \longrightarrow CO + y_{/2} H_2$$

The efficiency of a gasifier is dependent on the design of it, which in turn depends on the type of feedstock used. For a very high temperature and pressure gasifier, the efficiency could be higher than 90% in converting the chemical energy present in the solid fuel into the chemical energy of the syngas. The cost of a gasifier is quite high, and the cost increases rapidly with the increase in design efficiency.

It is to be noted here that the only direct use of syngas is as fuel; the CO, being highly toxic and corrosive, cannot be stored safely, especially because CO is a colorless and odorless fatal gas. However, it is a suitable feedstock for the petrochemicals industry.

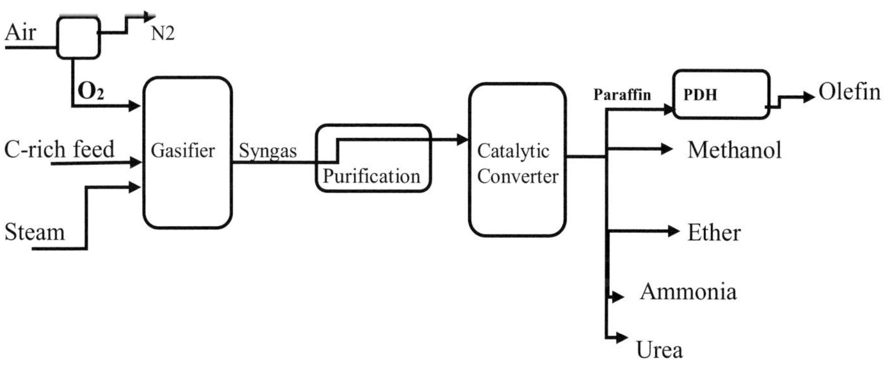

Fig. 6.5.6. Conversion of low-value carbon source into petrochemicals via gasification

Figure 6.5.6 shows a simple block flow diagram of an alternate process to convert a low-value carbon-rich source such as petcoke into value-added petrochemicals. The better the integration of refinery with these technologies and the petrochemical industries, the better the overall economics.

Synthesis of several varieties of desired molecules could be done from syngas simply by varying the ratios of CO to H_2, such as targeted hydrocarbons or oxygenated compounds like alcohols and ethers or ammonia or urea, etc.

Methane is produced from syngas at a high hydrogen to CO ratio in the presence of nickel-based catalyst according to the following reaction:

$$CO + 3H_2 \longrightarrow CH_4 + H_2O$$

Methanol (CH_3OH) was commercially produced from syngas by reacting over copper-zinc catalyst at around 250 °C–400 °C and 10 to16 MPa pressure (discussed in subsection 6.5.4). Now, due to the availability of abundant natural gas, this technology is not economical anymore.

Methanol is also used for the production of methyl-tertiary-butyl-ether, commercially known as MTBE. Methanol is reacted with an iso-olefin to produce MTBE, as follows:

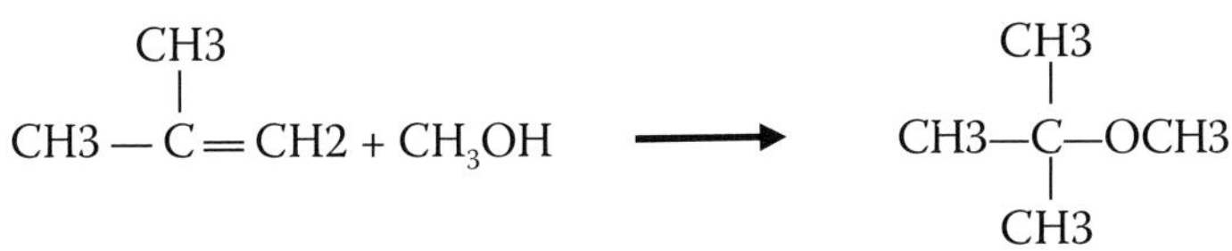

MTBE

This compound is used as a gasoline additive to improve the octane number. However, this product is now banned in several countries, including the US, for its toxicity and water contamination.

Dimethyl ether (CH_3 –O-CH_3) is also a product of syngas. Mobil developed a process called "methanol-to-gasoline" through dehydration of methanol to DME to gasoline over zeolite catalyst. The process has its own operational problems due to the formation of unwanted byproducts, especially waxes that create fouling.

Syngas can be further converted into additional hydrogen as follows:

$$CO + H_2O \longrightarrow CO_2 + H_2$$

In this commercial process, carbon monoxide is reacted with steam in the presence of a metal catalyst in a slightly exothermic process called a water-gas-shift (WGS) reaction.

6.6 Production of liquid hydrocarbons from syngas

Conversion of syngas to liquid hydrocarbons was developed before World War II by the Germans in their classic process well-known as Fischer-Tropsch (FT) synthesis. Several commercial plants were built in Germany during the war to supply fuel for their war machine. Because the process was very expensive, after the war the technology never got much attention. However, during the Apartheid period in South Africa, uncertainties about long-term oil supply and cost forced them to build FT plants to convert locally available coal into liquid hydrocarbons as a potentially viable option.

The major products from syngas are straight-chain synthetic hydrocarbons, where different types of hydrocarbons are produced depending on the ratio of CO to hydrogen at the required temperature and pressure over a catalyst. Depending on the type of product hydrocarbon desired, the FT process can be done at lower temperatures (200 °C–250 °C) or higher temperatures (300 °C–350 °C) with iron or cobalt-based catalyst. The reaction proceeds according to the following mechanisms:

$$nCO + (2n+1)H_2 \longrightarrow C_nH_{2n+2} + nH2O$$

$$2nCO + (n+1)H_2 \longrightarrow C_nH_{2n+2} + nCO_2$$

$$nCO + 2nH_2 \longrightarrow C_nH_{2n} + nH2O$$

Under the severe conditions when CO conversion is higher, hydrogenation of CO_2 takes place producing some additional paraffinic hydrocarbons, as follows:

$$nCO_2 + (3n+1)H_2 \longrightarrow C_nH_{2n+2} + 2nH_2O$$

In a medium-pressure FT synthesis, the major product (30-35%) is gasoline (C_{2-} C_{12})-range hydrocarbons and the next product (10-12%) is diesel (C_{12}–C_{20})-range hydrocarbons with a very high “Cetane” number (above 70), whereas the normal petroleum diesel is of Cetane number below 50. Usually C_2 to C_4 hydrocarbons are produced near 10% range.

Although the FT process can produce all ranges of hydrocarbons, by controlling process conditions of temperature and pressure and proper selection of catalyst, one can narrow the range of desired hydrocarbons produced. For example, maximum yield of gasoline and diesel transportation fuel can be obtained at 240–340 °C temperature range and at a medium pressure of 1.0 to 5.0 MPa, over an iron-based catalyst promoted by potassium oxide. The process is usually carried out in a fluidized bed reactor.

6.7 Manufacture of polyolefins/plastics (Partial source: Hydrocarbon Processing 2003)

The commercial polymer manufacturing process is not as simple as it sounds. The polymer manufacturer tries to control the polymerization process with a target polymer quality. For example, the polymer should have some specific macromolecular structure, with certain molecular mass and density. These can be achieved only through highly controlled reaction conditions at a certain level. However, note that in catalytic processes, either a fluidized bed reactor or stirred tank reactor is used for the manufacture of polyolefins.

Union Carbide built the first commercial plant for the production of plastics from gaseous hydrocarbons in the late 1960s. The plant was built for the manufacture of polyethylene from ethylene monomer as feedstock in a vapor-phase-polymerization fluidized-bed reactor. Then Union Carbide came up with their famous UNIPOL process for polyethylene and polypropylene plants; later on, the technology was successfully marketed all over the world by Dow and Exxon.

Fluidized-bed technology has since been upgraded by various other companies with their own proprietary information. The catalyst technology has also been improved since the discovery of the pioneering famous Ziegler-Natta catalyst ($TiCl_4$ / $MgCl_2$), and today's catalysts are tailored to the requirement of the target polymer product. Basically, it runs with the Ti-based catalyst supported on chromium carbonate, fluidized between 100-115 °C at a pressure between 15 to 25 atmospheres. During the polymerization process, a small variation of the feed rate or catalyst circulation rate or polymer withdrawal rate can cause hot spots formation and that will melt the plastics inside the reactor. Efficient control of the heat is essential to maintain the quality of the polymer. Some companies are also using a stirred tank batch reactor system with homogeneous catalyst.

Usually, polymers are characterized by their density and melt-flow index. For example, for the manufacture of polyethylene (PE) , there are in general two types of PE polymers: high density (HDPE) (>0.93) and low density (LDPE) (<0.93).

The density of the polymer decreases with the increase in molecular weight and branching of the side chains. The density of the polymer increases with the decrease in pressure and increase in temperature. Melt-flow index is indirectly a measurement for the chain length of the polymer and hence its viscosity. The higher the index value, the shorter the chain length. Both density and melt-flow index are controlled at the polymerization step at the interface of the gas and solid on the catalyst surface.

Polymerization being exothermic, control of the reaction temperature is quite difficult through internal quenching systems. The greatest challenge for the operators is to control the temperature with efficient removal of extra heat. If "hot-spots" occur inside the reactor over the catalyst surface, the fluidization pattern changes and polymerization starts at the hot-spot at a much higher rate, increasing the temperature further and melting the plastic. Once that starts, a chain reaction occurs and the only option is to shut down the plant.

A polymerization reaction is also highly dependent on the type of reactor used, whether it is a batch reactor or continuous reactor. For ethylene polymerization, the average heat of reaction is about 20 kcal /mol. Thus, a highly efficient cooling system is required to remove the heat from the reactor.

6.7.1 Thermal polymerization process

Figure 6.7.1 illustrates a schematic diagram for the production of LDPE in a non-catalytic thermal process. Thermal processing takes place in a tubular reactor at a high pressure of 1000–1500 atm and low temperature between 150 to 250 °C. Ethylene feedstock is compressed in stages to 1500 atmosphere before entering the tubular reactor. The inlet temperature of the reactor is maintained at around 200 °C, and the outlet temperature is cooled to around 120 °C. The outlet pressure is dropped in steps through pressure letdown valves to atmospheric pressure and temperature. Unreacted ethylene is recycled back. Polymer is withdrawn at the bottom of the separator with the help of an extruder.

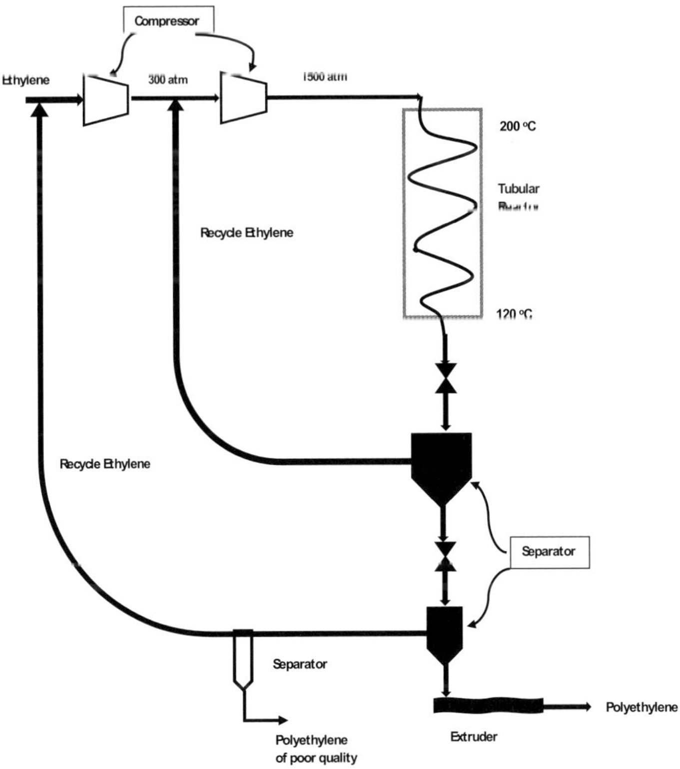

Fig. 6.7.1. Simplified schematic diagram of thermal processing of ethylene for the production of low-density polyethylene

6.7.2 BP process—hybrid technology

BP commercialized a hybrid technology for the polymerization of ethylene and propylene into plastics. Figure 6.7.2 represents a schematic diagram of the BP process (source: BP website).

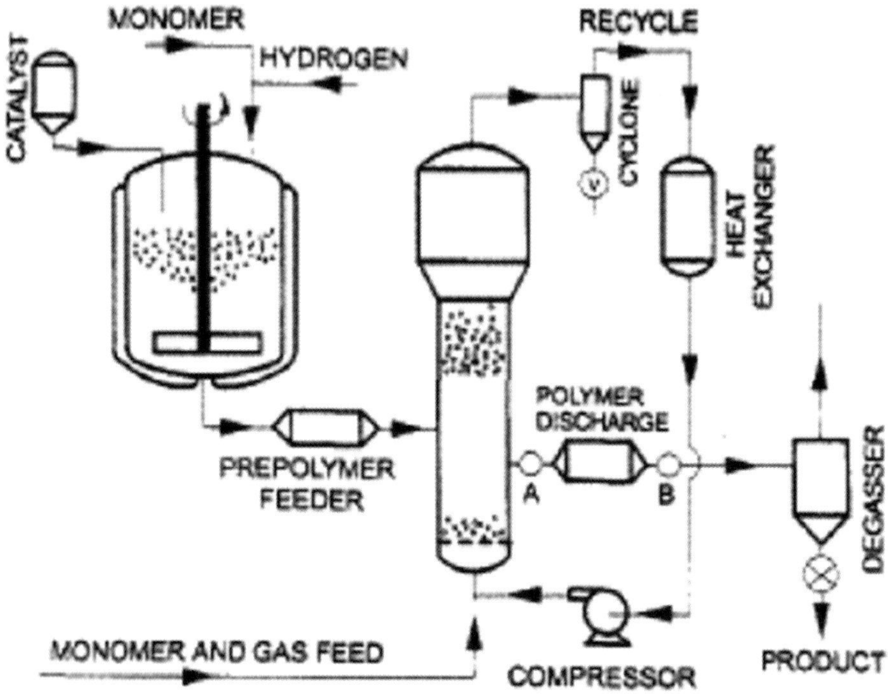

Fig. 6.7.2. Hybrid technology by BP

As shown, it is a combination of a slurry-tank batch process and a fluidized-bed continuous process in series. The fluidized-bed portion of the technology was originally developed in Russia and was very similar to the design developed by Union Carbide's UNIPOL process at that time in the US. BP modified the process with the addition of a pre-polymerization step in a slurry-tank batch reactor in front of the fluidized-bed reactor.

Catalyst, monomer, and hydrogen were added in an n-heptane solution inside the slurry-tank from the top, and the partially reacted active-monomer along with the catalyst is removed from the bottom and directly fed to the fluidized-bed reactor. The stirred tank usually runs between 50 °C to 125 °C, and the pre-polymerization is carried out in the liquid medium. This additional pre-step provides better control of the fluidized-bed operation and controls the polymer properties.

6.7.3 Olefin polymerization technologies by Axens

There are several processes available in the world and several companies producing olefin as intermediate product for the polymer manufacturing industries. However, there are very few process developers who provide a whole chain of process schemes from olefin production to polymerization process and finally making the finished product such as high-density or low-density polymers.

There are three major steps involved in the whole process scheme:

i. Manufacture of olefins
ii. Separation and purification to produce high-grade olefins
iii. Selective polymerization process to manufacture targeted polymers

Axens is one of the leaders in providing all the above steps for the plastic manufacturing industries. Axens claimed to have developed highly specialized adsorbents in order to obtain ultra-high-purity olefins such as ethylene, propylene, or butenes. Axens is also a provider of catalysts for the selective polymerization processes of olefins.

A list of Axens' extensive portfolio of olefin polymerization technologies is shown below (source: Axens website):

Feed		Process		Product
Ethyelene	→	Dimerization AlphaButol	→	Butene-1
Ethyelene	→	Trimerization AlphaHexol	→	Hexene-1
Ethyelene	→	Polymerization	→	Polyethylene LDPE / HDPE
Propylene	→	Dimerization Dimersol	→	Hexene
Propylene	→	Polymerization	→	Polypropylene
Butene	→	Dimerization Dimersol	→	Octene

Fig. 6.7.3. Axens polymerization technologies

References for Chapter 6

1. Ref: https://www.refiningandpetrochemicalsme.com/petrochemicals/ (Aug 2018).
2. V. Rao, *Shale Gas*, RTI publication (2012).
3. William Leffler, *Petroleum Refining 3rd Edition*, publisher PennWell (2000).
4. C. C. King , British Chemical Engineering, No 11, 32, (1959).

7

Thermal processing of heavier feedstocks

As crude oil has become heavier and heavier, there has been a substantial increase in residue upgrading and processing capacities worldwide. Thermal cracking processes continue to be the most popular choice for the high conversion of residue into lighter distillate product because of its low operating cost. There are two major thermal cracking processes:

(i) Deep thermal conversion—Coking process
(ii) Mild thermal conversion—Visbreaking process

Under the deep conversion processes, there are three major commercially available technologies:

(a) Delayed coking—the most popular commercialized technology
(b) Fluid coking/Flexi coking—developed by Exxon
(c) Eureka process—developed by Chiyoda Corp., Japan

The United States has more than 50% of the world's delayed coking capacity but very little visbreaking because of low demand for the residual fuel oil product, whereas it is quite popular in Europe.

7.1 Delayed coking process

To meet the growing demand for hydrocarbon resources in the transportation sector and in the petrochemical industries, it is necessary for refineries to employ low-cost heavier feedstocks in order to maximize the production of the target products. It is true that heavy crude is very difficult to handle, from recovery to transportation to refining. In order to upgrade the bottom-of-the-barrel, heaviest, and dirtiest of the heavy crude oil, many refiners employ delayed coking to convert the material into lighter product and to partially remove the impurities, like sulfur, nitrogen, metals, and asphaltenes, along with coke.

Delayed coking is one of the most severe thermal cracking processes with no catalyst added. In order to break the big polynuclear aromatic structures present in the heavier feedstocks, a higher temperature is required for a longer period of time, which is achieved in a delayed coker. A schematic diagram of the delayed coking process is shown in Figure 7.1.1. Preheated feedstock enters the bottom of the coker drum at around 490 °C–510 °C at a low pressure of 10 to 50 psig. Thermal cracking reactions start at the bottom of the drum and after a short lag time period, it starts producing two main products:

(i) the solid coke that remains at the bottom of the coke drum, and
(ii) a hydrocarbon cracked vapor product that is drawn off the top of the coke drum

The drum exit temperature at the top is around 430 °C to 450 °C, much lower than that of the feed at the bottom of the drum. As the layers of coke form, the hot feedstock passes through the porous hot coke bed and the thermal process repeats at the top of the coke bed. The coke bed height increases with gradual accumulation of coke until the drum is full. The overhead product stream is fed to a fractionator to separate the gaseous hydrocarbons and various fractions of liquid hydrocarbons.

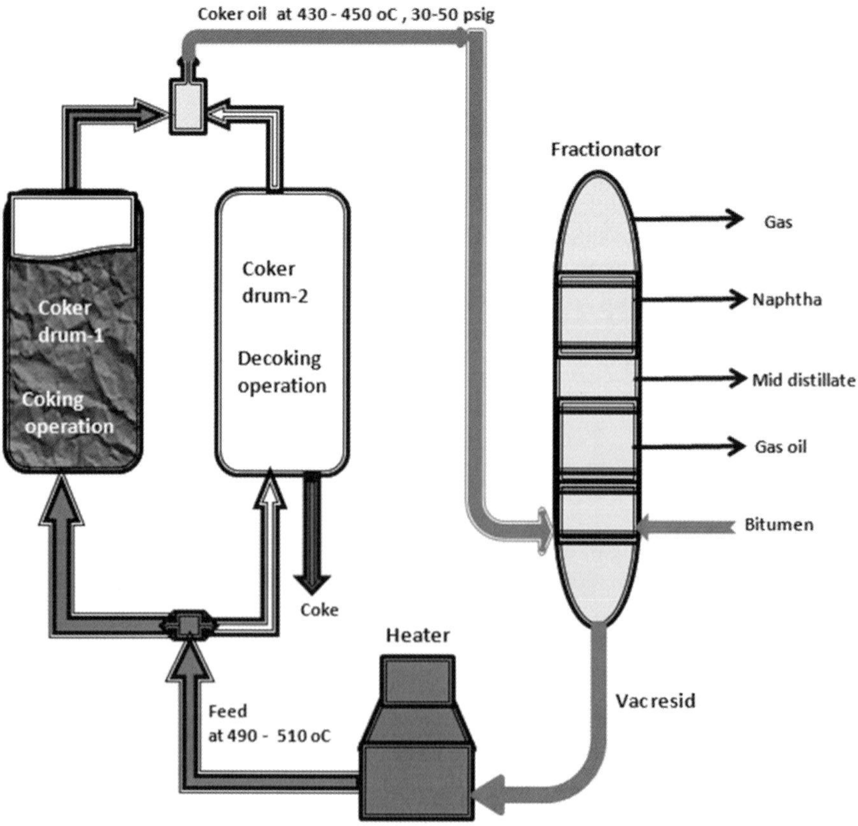

Fig. 7.1.1. A typical arrangement of a delayed coker plant (source: author reference 2, Chapter 1)

Due to its simplicity and cost-effectiveness, refiners widely use delayed coking to handle difficult-to-upgrade heavy feed. Because of carbon rejection and the formation of coke, the coker drum gets filled as the process proceeds. Thus, eventually it is a semi-batch process; when the coke drum is full, it needs to be switched to another one. As shown in the figure, in order to make it a continuous operation, there are always two coke drums in a commercial plant. In a refinery, typically it works in a 12-hour cycle; once one drum is full in 12 hours, another drum has been decoked and cleaned to get ready for the next operation. Decoking is carried out by drilling a vertical hole through the center of the coke and then the coke is cut with a high pressure water jet pumped through the hole. Water and coke drop to the bottom of the coke drum. Dry coke is then transported to a storage area.

There are various types of coke, graded depending on their quality, which is directly dependent on the quality of the feedstock itself. It is sold in the market based on the grade of the coke and valued accordingly; for example, shot coke is the poorest quality coke, used as fuel sold at the lowest price; sponge- or anode-grade coke is medium grade coke; and needle coke is the highest quality coke, sold at a very high premium price to the graphite manufacturing industry.

A typical feed for the delayed coking process is from the bottom of the vacuum distillation tower, known as vacuum resid, as shown in Figure 7.1.1. Usually, vacuum resid is a very high viscosity liquid, which is almost solid at room temperature. Its boiling point is above 1000 °F or 540 °C. Properties of a typical deep-cut vacuum resid from Canadian bitumen are shown in Table 7.1.1. The properties will vary depending on the source of the heavy crude oil and the temperature cut of the vacuum distillation tower.

Table 7.1.1. Properties of a typical vacuum resid

Properties , units	
API, degree	2.5
n-C5 - Asphaltenes, wt%	19.0
CCR, wt%	24.5
Sulfur, wt%	5.5
Nitrogen, wt%	0.6
Metals V / Ni, ppm	~800 / 200
Kinematic viscosity @100C, cSt	~100,000

Asphaltenes, by definition, is a material insoluble in normal paraffinic solvent, as in this case, the vacuum resid has about 20% of n-C5-asphaltenes, i.e., the material insoluble in n-pentane. The remaining 80% of the material is called maltenes, and in this case, mostly it is called resins.

For a commercial delayed coker operator, two important parameters responsible for the coke formation are asphaltenes content in the resid and its carbon residue value. One can predict the expected coke yield and hence the coker drum cycle run time from the Conradson Carbon Residue (CCR) or Micro-Carbon Residue (MCR) number of the feed. As a rule of thumb, weight percent coke yield is almost 1.5 times the CCR or MCR number. The amount of coke yield also depends on the asphaltenes content of the feedstock.

Figure 7.1.2 illustrates examples of coke yields depending on the CCR values, asphaltenes contents, and their asphaltenes/CCR ratios for four different vacuum resids. The coker feedstocks are from different sources that include conventional and unconventional crudes. Coke yield clearly increases with the increase in CCR values and asphaltenes content in the feedstock. It also follows the scale of asphaltenes to CCR ratios, where the coke yield increases with the increase in the ratio. However, results also show that when the CCR number is very high (>20%), it may not follow the rule of thumb, as the asphaltenes play an important role too. For feeds 2 and 3, the CCR values are almost the same (20%), but the coke yield is higher for the 3rd feedstock containing higher amounts of asphaltenes (20%). Results also show that the coking process is quite complex, too, and not always the same rule applies. As shown, in the case of the fourth feed with an asphaltenes / CCR ratio of 0.79, lower than the third feed (ratio 0.95), it produced more coke (31%), maybe because of higher CCR value.

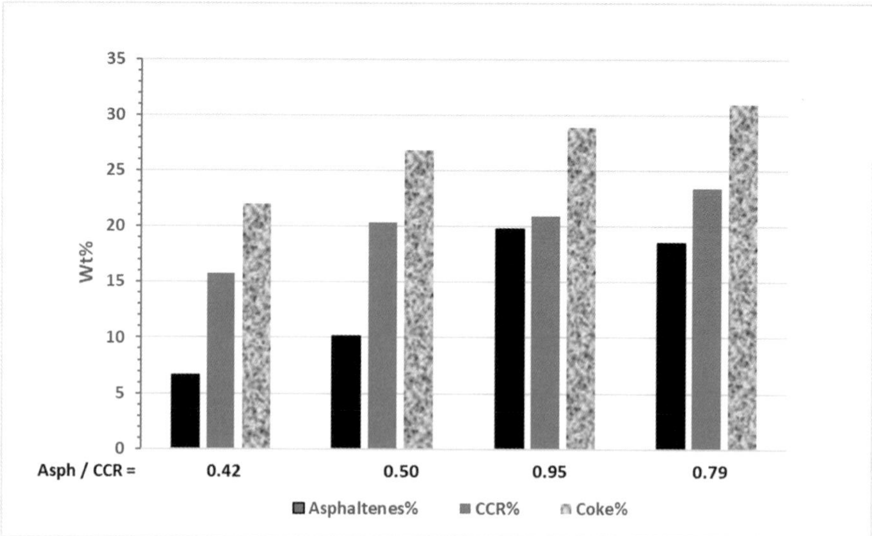

Fig. 7.1.2. Variation of coke yields with asphaltenes and CCR content of the feed

Figure 7.1.3 represents a volumetric liquid flow diagram of a typical delayed coker plant based on 10,000 BPD of vacuum resid feedstock obtained from the Canadian bitumen. As shown, the main liquid product is the heavier gas oil followed by the lighter naphtha and relatively smaller volume of diesel. The liquid volume ratios depend on the feedstock properties. About 25 weight percentage of the feed goes into the production of coke, and another 5 weight percentage makes up the gaseous hydrocarbons.

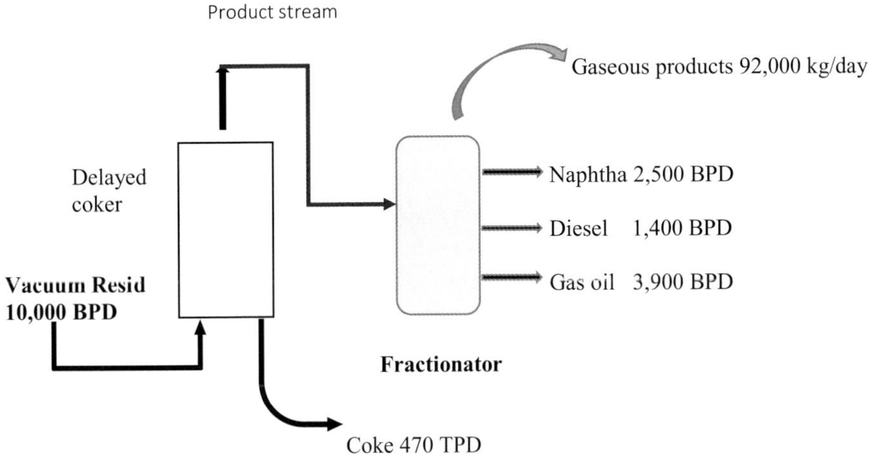

Fig. 7.1.3. Detailed product distribution from a delayed coker plant

The delayed coker may have other types of feedstocks such as atmospheric bottoms and the asphalt from a solvent deasphalting (SDA) plant. Depending on the refinery, in some places operators redirect decant oil from the FCC unit as a blend in the coker feedstock.

Due to the heavier feedstock used in a delayed coker, the liquid product obtained usually consists of higher concentrations of sulfur and nitrogen, the naphtha is highly olefinic, and the gas oil is highly aromatic in nature. Thus, all delayed coker plants must be accompanied by a higher-grade hydrotreater plant to meet the stringent environmental regulations.

7.1.1 Environmental issues with the delayed coking process

Petroleum refineries are complex systems of multiple processes that generate a significant amount of gases and wastewater contaminated with toxic hydrocarbons. Emissions from a commercial coker are difficult to quantify with a high degree of accuracy. Various types of hydrocarbons are produced, with variable toxicities. Measurement programs are time-consuming, costly, and also involve complex metering requirements.

The two most toxic materials are phenols and cyanides. They are produced in a delayed coker plant depending on the chemistry of the feedstock. Phenolic compounds are present in crude oil and end up in the wastewater streams from different processes in the refinery due to their high solubility in the water. The majority of the phenolic compounds in refinery wastewater originate from catalytic cracking and the delayed coking plants.

On the other hand, cyanide compounds are formed in the delayed coker unit at high severity coking conditions. Hydrogen cyanide (HCN) is formed by the combination of hydrogen and nitrogen present in the feedstock. Hence, the cyanide produced in coker wastewater may be free HCN and organic cyanides; they are not only toxic but also highly poisonous.

The cracking unit (catalytic or coking) vent is often the single largest emissions vent at the refinery. A single coke drum is typically on-line receiving feed and then off-line for cooling and decoking in a cycle of 12 to 14 hours. Near the end of the steaming process, a vent is opened to allow the remaining steam and vapors to be released into the atmosphere prior to draining and opening the drum. This atmospheric vent is commonly referred to as a steam vent:

- Steam vents are primarily methane and ethane, but include a variety of volatile and semi-volatile organic compounds
- Organic constituents account for the emissions from the entire decoking process operations:
    - venting, draining, drum opening, and coke cutting

The delayed coking unit is basically a closed system. When the coke drum is taken off-line, the initial steaming process gas is also recovered through the unit's product fractionator. The gas is typically sent to a flare, a flare gas recovery system, or the refinery's gas plant.

7.1.2 Foaming events in delayed coking

In processing heavy crude oil at higher temperatures, whether it is visbreaking or distillation or delayed coking, foaming problems can occur. In the delayed coking process, foaming occurs frequently and diminishes the efficiency of the process. It is a challenge for refiners to control the foam. In the case of severe foaming, the foam reaches into the overhead product and degrades the quality of the coker liquid. Some of the important factors responsible for foaming events are paraffinic feedstock or condensate present, production of lighter paraffinic hydrocarbons, pressure swing inside the live coke drum, colder temperature at the top of the vapor exit point, or injection of colder feed at the beginning of the empty coke drum, etc.

Commercially, the most widely used antifoaming agent is a silicon-based organic compound called Polydimethyl Siloxane (PDMS). It is used for economic reasons because of its low dose use and effectiveness. But it encounters other expensive problems in the downstream refining. PDMS tends to break down at higher temperatures inside the coke drum, disperse into the vapor phase, and the silicon reaches the overhead product line. In particular, the silicon is a problem for the downstream hydrotreater plant, as it poisons the catalyst.

In commercial operations, the antifoaming (AF) agent is injected through a nozzle from the top of the drum, where temperature is lower (800-850F) than the bottom of the drum (900-930F). The AF injection rate varies depending on the foaming events and foam layer height in the drum. The AF injection preferably begins when the drum is more than half-full, such that the AF agent reaches the foam layer prior to being cracked or vaporized and swept out of the drum. In recent years, refiners are switching to high viscosity (60,000 to 100,000 cSt) silicone-based antifoam agents as those are more thermally stable at coking conditions. However, recent research at the University of Tulsa in Oklahoma has shown that aromatic material such as tire oil is quite effective as an antifoaming agent[1].

Figure 7.1.4 shows the variation of densities with the type of products inside the coke drum, and it shows that the foam is always in the vapor phase above the liquid layer and the coke layer. Foaming events inside the live coke drum could be detected with the help of a gamma-densitometer, and its density varies between 200–300 kg/m^3.

7.1.3 Most plausible reaction mechanism of the delayed coking process (Ref 6)

The vacuum residue is a very complex mixture of hydrocarbons. As stated by Speight in his book, asphaltenes are a polymer of poly-condensed aromatic and naphthenic structures with long aliphatic side chains with heteroatom in it[2].

Based on several experimental studies carried out by the author over the years, a new thermal cracking reaction mechanism of vacuum residue has been proposed in this book for the first time. The proposal is based on the step-wise phase transformation that takes place inside the coke drum with the formation of vapor and coke products. The coke formation takes place through intermediate "mesophase" formation during the initial stage of the thermal cracking process. The mesophase is the non-vaporized product formed due to poly-condensation of the polynuclear aromatic hydrocarbons at the coking conditions that remain inside the coker drum as pre-cursor to coke formation.

Figure 7.1.4 illustrates what happens to the feedstocks as the hot feed enters from the furnace at the bottom of the coke drum. The phase transformation can be monitored visually in real time if someone is equipped with a gamma-densitometer that is capable of measuring the density inside the live coke drum. The cracking reaction starts, and it takes time for the maximum cracking to take place; then a part of the cracked product vaporizes at the coker inlet temperature (490–500 °C) and exits from the top of the drum. Another part remains as mesophase pitch at the bottom of the drum, and the reaction proceeds in the liquid phase. It is to be noted here that only in the liquid phase polymerization reactions take place with the increasing molecular weight (see explanation in subsection 3.9). As the phase transformation takes place, the bulk density increases from the cracked liquid (> 300 kg/m^3) to the mesophase (~700 kg/m^3) to the coke (~900 kg/m^3) over time, as shown in Figure 7.1.4. Coke formation is basically a carbonization reaction where carbon residue remains with the elimination of hydrogen from the aromatic nucleus, as explained in detail about the carbonization process in section 3.6 in Figure 3.6 and Table 3.6.1.

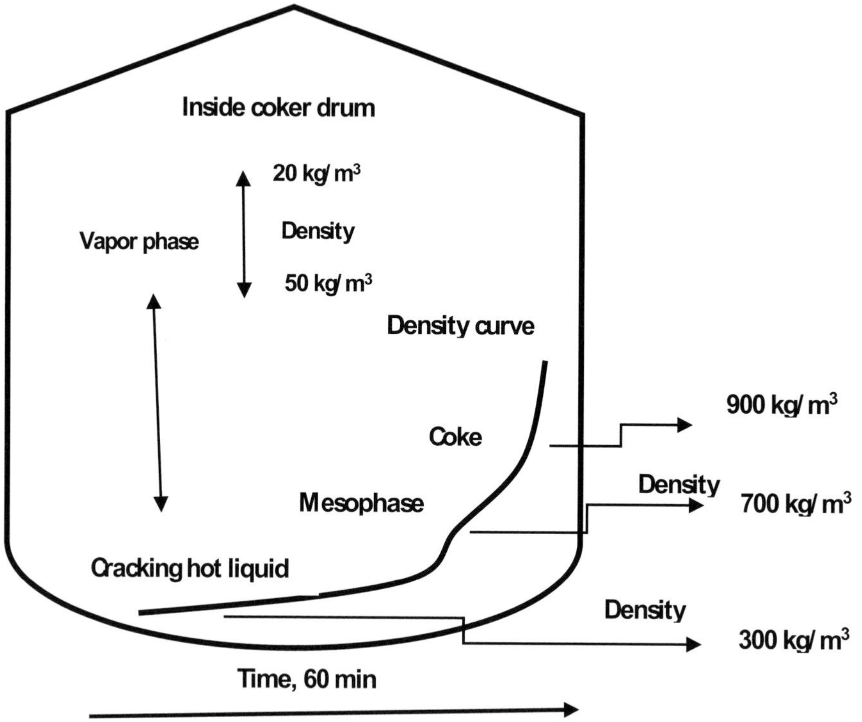

Fig. 7.1.4. Variation of phase transformations and the density changes inside the coke drum as a function of time

The total process takes place over a period of 45–60 min before the whole delayed coking process reaches a steady state. The cracking process being endothermic, a significant amount of heat and time are required before complete cracking takes place and mesophase turns into coke. This is the reason why it is called delayed coking, where coke formation is delayed inside the coke drum.

One can also observe how the reaction proceeds over time before the steady state is reached through real-time on-line gas analysis and liquid analysis in short time intervals. Once the steady state is reached, the rate of coke formation and the product distribution remains the same, and there is no more change in the liquid or gaseous composition. The process repeats with layer after layer of coke formation inside the drum, and the coke height increases. In a commercial delayed coker plant, the coke drum is equipped with a gamma-densitometer at the top that can detect the foaming event and also when the coke drum is filled with the coke, in order to switch to the second coke drum.

The most important data used for the proposed reaction mechanism is the carbon number distributions in the coker liquid products. Figure 7.1.5 illustrates the carbon number distributions of various sub-products from the delayed coking of two different types of feedstocks at 930 °F, i.e., in one case, the resid is from a conventional heavy crude oil and in another case from an unconventional heavy crude oil. The sub-product yields are almost of the same order from two different sources of vacuum resids.

The carbon ranges represented here were determined from the simulated distillation curve of the coker oil. The carbon ranges were divided based on the boiling points of their corresponding equivalent paraffinic carbon numbers. But strictly speaking for higher molecular weight (>C30), the boiling points of aromatic hydrocarbons do not fall in the same carbon range as those of the non-aromatic hydrocarbons.

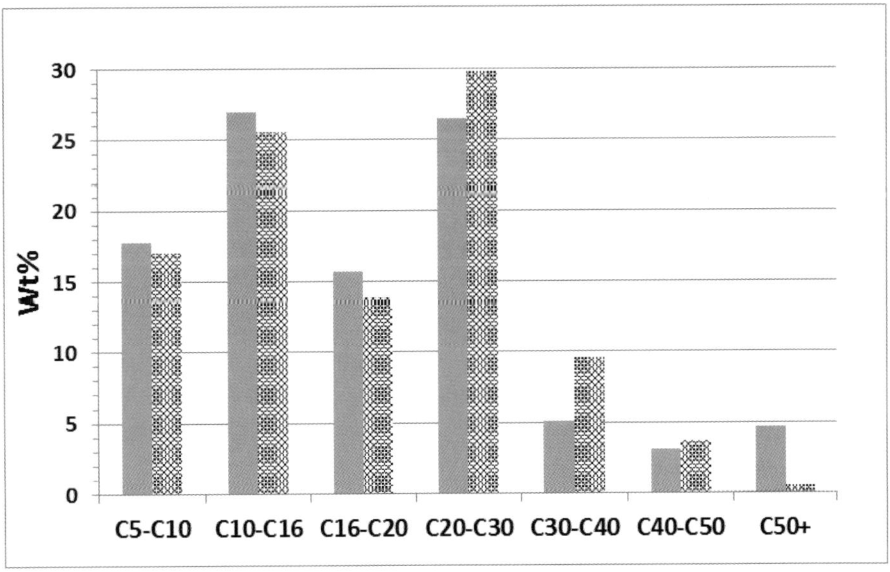

Fig. 7.1.5. Carbon number distribution in coker liquid from the vacuum resids of unconventional (solid color) and conventional (dotted) crude oils

Figure 7.1.5 clearly illustrates that in the liquid product, there is a relatively high concentration (>25%) of C20–C30-range (LGO—light gas oil) and C10–C16-range (kerosene) hydrocarbons. However, for the conventional feed, the C_{20}–C_{30}-range hydrocarbons are at a higher concentration (at 30%) than the

C_{10}–C_{16}-range hydrocarbons (at 25%), whereas for the unconventional feed, both were almost of the same amount (~ 27%).

Next in line is the C_5–C_{10}-range naphtha fraction (just above 15%), which is at a higher concentration than that of the C_{16}–C_{20}-range diesel fraction (just below 15%). There are quite small amounts (< 5%) of hydrocarbons higher than C_{40} range (HGO—heavy gas oil) and C50+ (pitch) in the coker liquid.

It is well-established that during the delayed coking process, random thermal cracking of hydrocarbons takes place with the formation of free radicals. Nevertheless, the present study shows evidence that when bigger size (>C40) molecules in the resid are subject to severe cracking, such as in delayed coking, a certain type of cracking reaction is preferred and dominates over others. The proposed theory is that whether it is a thermal process or a catalytic process, certain types of products prevail, indicating that a certain type of reaction is likely to occur and contribute to the process.

Understanding the molecular structure composition of the coker feedstock is crucial in order to hypothesize any thermal cracking mechanism. It is known that heavy crude oil or bitumen contains a large quantity of poly-nuclear aromatics and beyond pyrene-type structure that makes it such a difficult feedstock to handle[2].

It is important to keep in mind here that the vacuum residue (VR) is the product of the vacuum distillation; the residue remains undistilled and consists of hydrocarbons of boiling point higher than 540°C (or 1000F+). Hence, a large majority of the hydrocarbons present, being poly-nuclear-aromatic in nature, are greater than the C40 equivalent paraffinic hydrocarbons. Thus, the coker liquid produced during the delayed coking process are cracked product consisting of molecules smaller than the feedstock (<C40).

Because of the complex structure of the coker feedstock, its behavior during the thermal cracking process is also very complex. In order to understand the thermal behavior, a plausible simplified molecular orientation of the VR is shown in Figure 1.4 in Chapter 1. It has been postulated here that the VR consists of three layers of structures, and it has been established that the core hydrocarbon structure of the VR is either exclusively or predominantly made of poly-nuclear aromatic molecules (shown in black aromatic structure) that are the most difficult to crack under the coking conditions.

The second inner layer of molecules represents the more dispersed aromatic-naphthenic type groups of maximum 3 to 4 ring structures. Resins are smaller in molecular size than that of the asphaltenes.

The third outer layer consist of mostly saturated hydrocarbons (> C40) and smaller ring aromatics of maximum 2 to 3 rings (> C30). This layer mostly represents the saturates and aromatics fractions of the SARA analysis. Those are also the long side chains extended from the inner layers of asphaltenes and resins. R_1 and R_2 represent aliphatic side chains of varying lengths. The third layer is mostly responsible for the formation of lower molecular weight liquid hydrocarbons.

As discussed below, the three layers may undergo thermal cracking reactions at different rates and at different times. The proposed thermal reaction mechanism is illustrated in Figure 7.1.6. The numbers shown are the average wt% yields of various ranges of hydrocarbons from a typical vacuum resid of conventional or unconventional sources. The compounds may split into various forms of lower molecular weight liquid or gaseous hydrocarbons, or the aromatics may polymerize to solid coke formation at the same rate or different rates under the coking conditions.

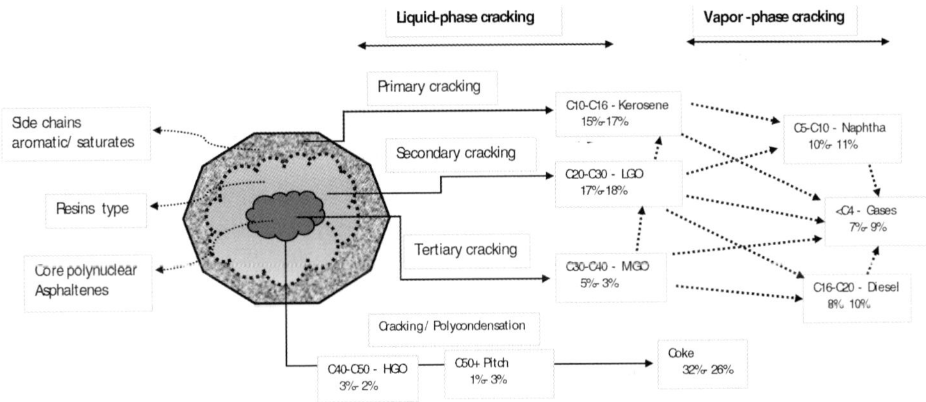

Fig. 7.1.6. Most plausible reaction mechanism of delayed coking process

Both the coker liquids from different sources have the highest yields (>15%) of kerosene (C10–C16) and LGO (C20–C30)-range hydrocarbons. The most plausible reason is that the higher yields of kerosene-range hydrocarbons are mainly due to the primary cracking in the liquid phase of the outer layer consisting of the side chains of the asphaltenes and resins. This is followed by further liquid phase cracking of the second inner layer consisting of aromatic-naphthenic structures into LGO-range hydrocarbons. It could be postulated here that during the liquid phase thermal cracking of VR, the molecular groups present in the second layer of the VR tend to break almost at the middle of the structure, i.e., the alkyl bridge between two aromatic-naphthenic nuclei, where the weakest C-C bonds are present, producing two or more larger-size molecules in the range of LGO.

Another new theory proposed in this book is that the higher yields of naphtha fractions (C5–C10, >10%) are due to the vapor phase secondary cracking of the kerosene and LGO hydrocarbons already produced from the initial liquid phase cracking of the VR. Similarly, the relatively lower yields of the C16–C20 diesel fraction is due to the fact that it is produced from the vapor phase cracking of the heavier >C30 molecules in the LGO and MGO-range hydrocarbons.

The vapor phase cracking follows a different rule in that it breaks near the end of the chain length, hence producing more of the lighter gaseous (<C4)-range hydrocarbons. Other possible reactions that take place randomly in the vapor phase are isomerization, free radical capping, dehydrogenation, and cyclization, while the cracking reaction dominates in the liquid phase.

The above theory explains the reason why yields of gaseous hydrocarbons (<C4) are so high (just below10%), as every fraction in the vapor phase in turn breaks down into lower molecular weight hydrocarbons. Due to shorter residence time in the vapor phase, the probability of cracking and dehydrogenation reactions is greater than the recombination reactions to form bigger molecules. This explains the reason why the diesel fraction is lower than the naphtha- and kerosene-range hydrocarbons in a delayed coker, as this is most probably a vapor phase product.

Fig 7.1.6 also shows that the yields of heavier >C30 hydrocarbons are quite low, below 5%, because they are formed from the liquid phase cracking of the core layer of poly-condensed aromatic structure after the outer and central layers are gone. The differences in the yields from two VR feedstocks are due to the difference in the core structure of the resids. As was noted previously, higher coke yield is due to higher concentrations of core structure, and hence the liquid phase cracking reaction dominates to produce more C30–C40 (MGO)-range hydrocarbons from the >C40-range hydrocarbons.

However, the resid's much lower yields (<3%) of C40–C50 (HGO) and pitch (C50+) material, further justifying the fact that the core structure of resid consists of more fused-ring aromatics that favors coking reactions and hence produced more coke (32%) and less C50+ pitch. This is consistent with the generally accepted theory that thermal cracking of the polycyclic aromatics produces free radicals (those fractions are also nonvolatile at coking conditions) that stabilize through recombination to form coke, similar to the step 3-termination reactions shown in Chapter 1.

Another study by the author[5] confirms further the above theory that the heavier liquids are the product of liquid phase reaction. In that study it was demonstrated that when the delayed coking was carried out in the presence of recycled lower hydrocarbons (C1-C4) (a by-product of the process) by injecting at the bottom of the coke drum (5 to 20%), there was an increase in the liquid product yield by 3 to 5 weight percentage point. The liquid product analysis showed that the increase was mostly in the heavier end due to stripping and cracking of the non-volatile liquid and capping of the free radicals with the gaseous hydrocarbons.

7.2 Fluid coking / Flexi coking

Delayed coking is the most economical process; it is widely used commercially to handle highly contaminated heavy hydrocarbons and to convert into lighter product. However, the biggest problem with the process is that it is a semi-batch plant. Hence, Exxon developed a continuous coking process utilizing a fluidized-bed concept, where coke particles are fluidized at a temperature between 500 °C to 550°C. As the coke forms around those particles, it is removed from the main reactor to a different heater where part of the coke is burnt. The hot coke particles remaining are then recycled back to the fluidized bed reactor as shown in Figure 7.2. Burning of the coke at the same time provides heat for the fluid coker. This fluid coker runs at a higher temperature than the delayed coker, and for a shorter reaction time. This helps in reducing olefin formation and further polymerization reactions towards coke formation, thus producing higher liquid yield and less coke yield, as compared to the delayed coker. Due to the burning of part of the coke, the effective net coke yield is much less than the delayed coker. The yields of products cannot be compared directly as the two processes handle feed and product differently—in general, 2 to 3 volume percentage points extra liquid yield and about 75% less coke yield from fluid coking than delayed coking.

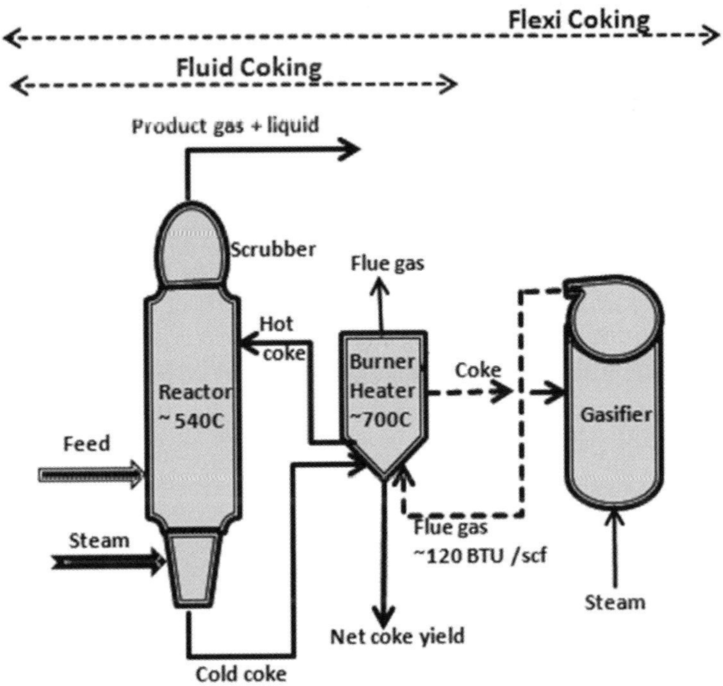

Fig. 7.2. Simplified diagram of fluid coking/flexi coking (source:author)

The coke in the fluid coking is extremely poor quality, very hard with high percentages of metals and heteroatoms. Its market value is poor; it can be used for cement kilns or to burn in a steam boiler. On the other hand, the delayed coker has the advantage of making better quality anode- or needle-grade coke, which a fluid coker can never produce.

The biggest disadvantage of fluid coking is that when high sulfur / nitrogen coke is burnt, it emits NOx/Sox, and also burning coke produces CO and CO_2. In order to reduce the emission of toxic gases, Exxon further modified the fluid coking process by introducing a gasifier after the coke burner, where the coke particles from the burner are gasified in the presence of steam, as shown in Figure 7.2. The improved process is called Flexi-coking. Thus, basically there is no fundamental change in the actual process; hence the liquid yield remains the same in the flexi-coking as that of fluid-coking at a higher cost of running the gasifier.

7.3 Eureka process by Chiyoda Corporation, Japan

The Eureka process was developed by Chiyoda Corporation and ran commercially in the Fiji Oil refinery in Japan. The main difference between delayed coking and the Eureka process is that Eureka does not produce any coke; rather, it produces a high melting point liquid similar to pitch. Also, the Eureka process is designed to produce higher liquid yield than that of a delayed coker by continuous stripping of the liquid product with steam.

As shown in Figure 7.3, the Eureka process uses a swing reactor similar to the two drums used in the delayed coking process. However, in this case, pitch is withdrawn, not coke, from one reactor while another one is in the operation cycle. The cycle time in Eureka is much shorter (3–4 hours) than that of the delayed coking (10–12 hours). However, the main difference is that super-heated steam is added from the bottom of the reactor to provide additional heat, reduce hydrocarbon partial pressure, and induce stripping effect inside the reactor in order to reduce over-cracking of the liquid product, thus enhancing the liquid yield.

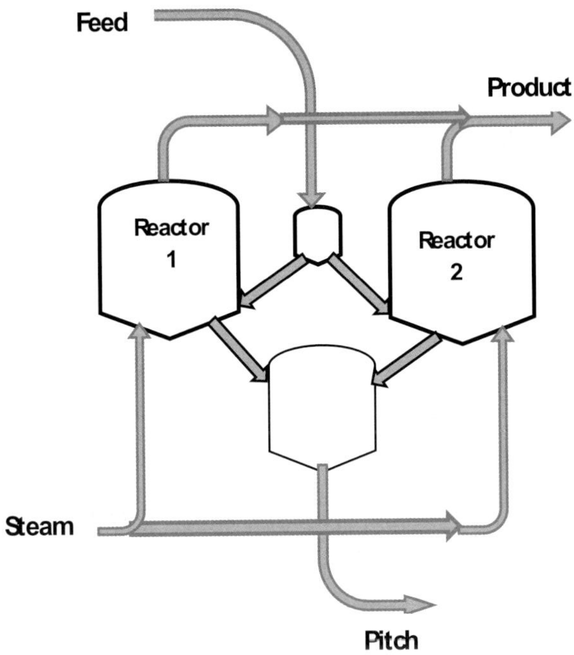

Fig. 7.3. Simplified diagram of the Eureka process

7.4 Deep catalytic cracking

This is basically a thermal cracking process carried out at a high temperature but in the presence of catalyst. The purpose of a catalyst is to reduce the temperature required to crack the same molecule without catalyst. In other words, one spends less energy to get the same product yield. Acid-based catalyst like zeolite promotes the cracking reaction and is widely used in refineries for FCC plant. Zeolites are made of alumina-silicate that leads to the formation of carbonium ions that promote cracking. In order to increase the olefin formation, as needed in the petrochemical industries, metal oxides impregnated with zeolites are used as catalyst to promote dehydrogenation and cracking. Propylene is the major olefin product in this process, and propylene production increases by almost 15% as compared to the thermal process. The catalytic process is good for the utilization of low-cost heavier feedstocks; however, the catalysts are prone to poisoning from the impurities in the feed and also deactivation due to coking.

Figure 7.4 shows a typical design of a deep catalytic cracking process to treat resid or bitumen. The design varies from company to company, as per their requirements. Cracking of heavy hydrocarbon is prone to coke formation on the surface of the catalyst. In order to maintain the catalyst activity and keep the process in continuous mode, it is necessary to have the catalyst replaced with fresh catalyst. In a fixed-bed catalytic process, it is difficult to replace the catalyst without expensive shutdown of the plant. Thus, it may not be economical to run a heavy feedstock through a fixed-bed reactor. Another alternative is to have the option of continuous replacement of the catalyst with a regenerated one. This is commercially done in other processes described earlier, as in the case of the FCC plant with some modification in operation.

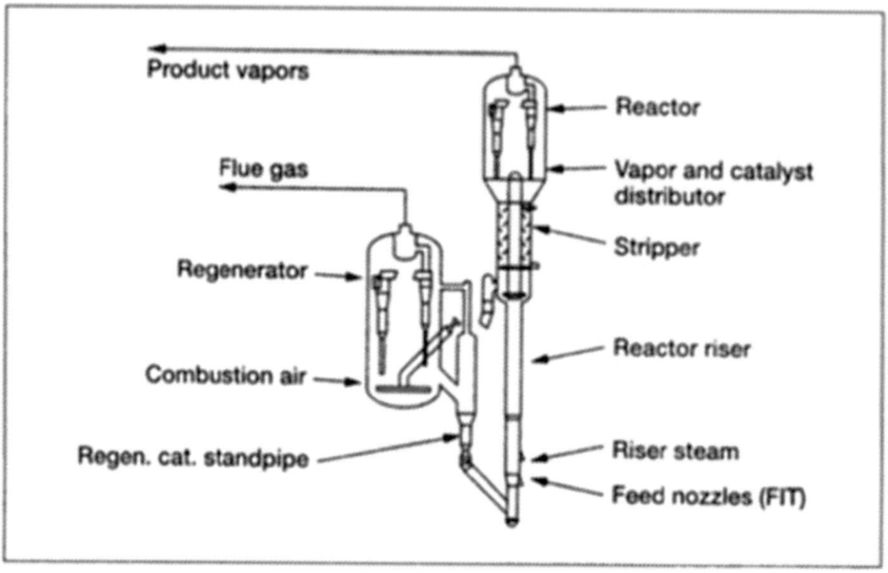

Fig. 7.4. A typical deep catalytic cracking process

In a deep catalytic cracking process (or resid FCC), the feed is atmospheric resid or equivalent diluted resid with limited amounts of Conradson carbon (CCR) and metal content to handle the coke deposit and the catalyst regeneration capacity. The vaporized heavy hydrocarbons come in contact with the catalyst particles in a riser reactor moving in upward direction. As the feed and the catalyst move together, the cracking reaction progresses. Heavier molecules break into smaller molecules and also form coke on the surface of the catalyst particles. The catalyst and hydrocarbon vapor are separated at the top of the reactor in the stripper section with the help of steam. The hydrocarbon oil is directed towards a fractionator to separate the oil into desirable fractions.

As shown in Figure 7.4, the catalyst is sent to a regenerator where it is regenerated by burning the coke on the catalyst surface with combustion air. The temperature of the regenerator is controlled to avoid overheating. The hot regenerated catalyst is separated from the flue gas in a cyclone inside the regenerator. The hot catalyst is recycled back and brought in contact with the feed hydrocarbons again.

The type of cracking catalyst will depend on the scope of the process. A refiner may like to crack paraffinic hydrocarbons or olefinic hydrocarbons. In other cases, a refiner may like to crack naphthenic or aromatic hydrocarbons. The purpose of the cracking is to produce a marketable product. For an optimum operation, the type of catalyst is to be tailor-made for a typical reaction condition and feedstock.

The most common commercial catalysts are:

i. Natural aluminosilicates, usually acid-treated to enhance cracking properties
ii. Synthetic silica-alumina—amorphous or crystalline—widely known as zeolites or molecular sieves.

Zeolite is the most commonly used catalyst because of its higher activity and lower rate of deactivation, increasing desired product yield at a relatively lower severity to avoid secondary cracking. Crystalline zeolites are more active than the amorphous type, but are more brittle and subject to attrition during circulation. Amorphous zeolite is less expensive than the crystalline one. Thus, in a commercial cracking operation, usually refiners use a mixture of both to achieve their target conversion. Approximately 15% to 20% of the crystalline provides the necessary target yield and at the same time makes it economically attractive. The mixture helps reduce the emission of catalyst particulates due to attrition and also reduces the overall deactivation rate.

It is important here to bring to the reader's notice that the deactivation of cracking catalyst is not taking place just because of coke formation on the surface, but the major problem is metal deposit. Heavy crude oil consists of nickel and vanadium as organo-metallic porphyrin structures and other inorganic metals such as iron, copper, calcium, and magnesium, etc. The deposit of these metals make the catalyst deactivated. Pore diameter plays an important role in processing of the heavy crude oil. Heavy hydrocarbons diffuse through the pore diameter into the pore volume for cracking to occur. Once the metals are deposited, it blocks the pore volume and cannot be removed simply by burning. Catalyst manufacturers make zeolite with designed pore size to handle the high metal content heavy crude oil.

7.5 Viscosity breaking (visbreaking)

Visbreaking is the oldest thermal cracking process commercially available for more than a century. The objective of the process is to break the long chain hydrocarbons to shorter ones in order to reduce the viscosity and pour point of the product material. Before visbreaking, heavy residue had little value and was used as residual fuel oil only. When the demand for lighter oil increased, this process was developed to thermally crack the heavier residue into lighter product. Then the technology was further developed to reduce the viscosity of the waxy material, which is a mixture of long chain paraffinic and aromatic molecules having high viscosity and pour point. Especially in winter, the waxy material cannot be transported.

With increasing demand for heavy crude oil, the visbreaking technology got more attention and was further extended to treat heavy atmospheric or vacuum bottoms to meet the viscosity specification for transportation purposes and to produce fuel oil. The demand for fuel oil in the United States is low, and visbreaking does not make any economic sense. In the US, the heavy residues atmospheric bottoms or vacuum bottoms are treated in a delayed coking plant.

There are two major types of commercial visbreakers:

(i) Coil visbreaker

(ii) Soaker visbreaker

As shown in Figure 7.5, both types of visbreaking plants, coil or soaker, are almost the same. In the case of coil visbreaking, the outlet of the furnace tube is cooled and the product stream is directly sent to a fractionator. By comparison, in the case of soaker visbreaking, a soaker drum is simply added in between the furnace and the fractionator. The purpose of the soaker drum is to increase the residence time and hence increase the degree of visbreaking. The critical parameters are temperature and residence time to obtain the desired degree of cracking. In the case of coil visbreaking, the temperature of the coil inside the furnace is maintained around 880 °F to 900 °F (470 °C–490 °C) and residence time of 1 to 2 minutes. In the case of the soaker drum visbreaker, the drum temperature is lowered and maintained between 820 °F–840 °F with residence times of 4 to 6 minutes. In the soaker drum, the pressure is also maintained high around 700-750 psig, in order to decrease the vaporization of the feedstock for liquid phase visbreaking. The main objective is to avoid a coking reaction in the soaker drum. Steam is injected inside the coil to increase the velocity and reduce the time in order to reduce the coke formation or fouling. The cracked product after reaching the desired conversion leaves the soaker drum, is quenched, and fed to a fractionator.

Residue conversion is higher in the soaker drum process. However, visbreaking is notorious for producing unstable product if the conversion level is increased beyond a certain point. Stability issues are more prominent for heavy crude and bitumen feedstock that contains asphaltenes. Visbreaking tend to increase asphaltenes precipitation, and the product becomes unstable. Increasing the conversion also increases sediment formation.

The stability of the visbroken product is determined by different methods by different companies and sometimes by their proprietary method. The degree of stability is not only dependent on the run conditions, but is a function of various other factors such as properties of the feedstock, degree of conversion, and the type of process used.

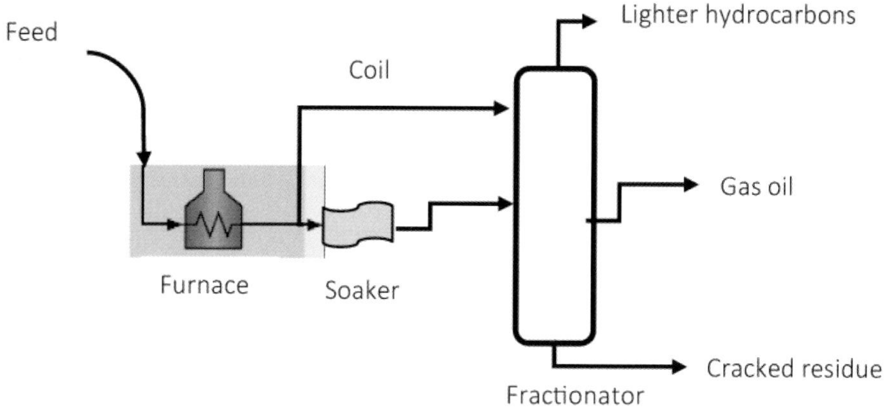

Fig. 7.5. Schematic diagram of two visbreaking processes

Figure 7.6 shows in an arbitrary scale the variations on the stability of two different feeds with the degree of conversion. It has been reported that the aromatic-naphthanic feeds are more stable for a given conversion, as compared to a paraffinic feedstock[4]. Thus, in a commercial operation, it is possible to obtain a stable visbroken product at a higher conversion level from an aromatic feedstock. Therefore, for a given feedstock and operating conditions, a coil visbreaker will produce more stable product than that of a soaker visbreaker.

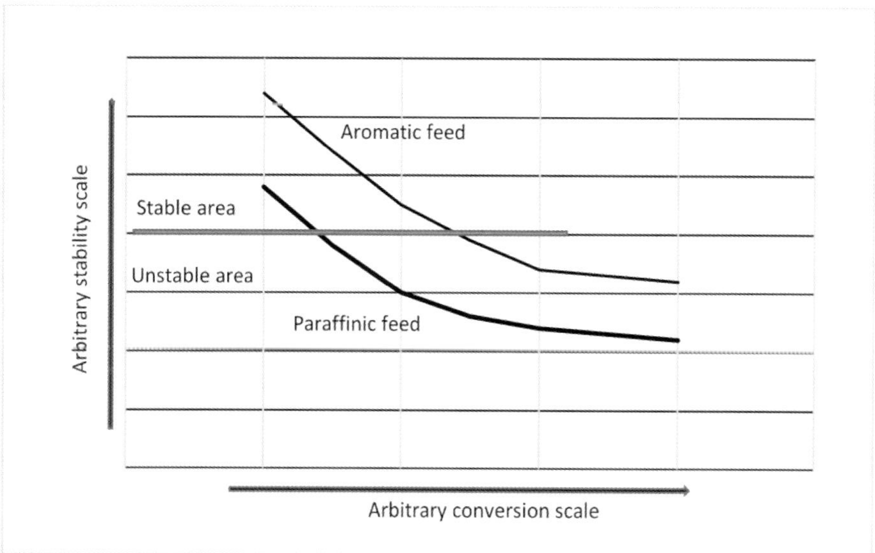

Fig. 7.6. Variation of stability of visbreaking product with degree of conversion

Visbreaking has a potential application as a pre-treatment of feedstock before a major process such as delayed coking. Visbreaking improves the quality of the feedstock and hence decreases the coke yield and increases the liquid yield in a coking process. Pretreatment of vacuum residue or high asphaltenes feed with visbreaking improves the quality of the coke morphology. However, constructing a dedicated visbreaker just for a delayed coker may not be an economical solution in all cases.

There are several visbreaking technologies available; among those, some prominent commercially available technologies are:

(i) Deep Thermal Conversion by Shell
(ii) High Conversion Soaker Technology by Toyo Engineering
(iii) Aquaconversion developed by Pdvsa, Venezuela.

Aquaconversion is basically a hydro-visbreaking process that is carried out in a catalytic slurry phase reactor in the presence of steam.

Readers interested in the details of these technologies may consult the original sources.

7.6 Gasification

The process of gasification is simply a severe thermal cracking with 100% conversion of heavy hydrocarbon residue or carbon rich petcoke or coal into gaseous products. The gasification is carried out at a temperature higher than 1000 °C, with a major product being syngas—a mixture of carbon monoxide and hydrogen along with methane and carbon black. The process did not get high attention because of its high cost. However, integrated gasification combined cycle (IGCC) is an attractive process for the power generation sector, as this emits a minimum amount of toxic gases such as SOx / NOx and CO_2 as compared to the present power generation plants.

The various chemical reactions that take place during gasification of carbon-rich material are as follows:

i. Combustion $C + O_2 \longrightarrow CO_2$

ii. Partial oxidation $C + \frac{1}{2} O_2 \longrightarrow CO$

iii. Boudouard reaction $C + CO_2 \longrightarrow 2CO$

iv. Steam reforming $C + H_2O \longrightarrow CO + H_2$

v. Shift reaction $CO + H_2O \longrightarrow CO_2 + H_2$

vi. Methanation at >1000 °C $C + 2H_2 \longrightarrow CH_4$

Other minor reactions that take place because of the presence of impurities in the feedstocks are as follows:

a. $S + O_2 \longrightarrow SO_2$

b. $SO_2 + 3H_2 \longrightarrow H_2S + 2H_2O$

c. $C + 2S \longrightarrow CS_2$

d. $CO + S \longrightarrow COS$

e. $N_2 + 3H_2 \longrightarrow 2\,NH_3$

f. $N_2 + xO_2 \longrightarrow 2NOx$

The product gas composition is obtained depending on the process condition applied.

The thermodynamics and kinetics aspects of each reaction drive the process and dictate how the gasification should be carried out.

In order to produce the best quality syngas ($CO + H_2$), the available options for the gasification of carbon are either through:

- Maintaining the steam to carbon ratio to carry out the steam reforming reaction

  or

- Carrying out partial oxidation of hydrocarbon in the presence of controlled starved oxygen

If the gasification is carried out with pure oxygen, the benefit is that at the end one gets nitrogen free synthesis gas (syngas), which is pure enough to be used for the production of petrochemicals as described in an earlier chapter.

The gasification process went through several generations of improvement since its first-generation commercial gasifier built by Lurgi after WWII to convert coal into syngas[3]. It is a vertical reactor; coal is fed through a hopper from the top, steam and oxygen injected from the bottom. Gasification is conducted at around 1000 °C and 3.0 MPa pressure. Because of the pressure, methane is favored during the reactions and major products are 50% hydrogen, 35% CO, and 15% methane.

The second-generation gasifier was developed by Texaco; a schematic diagram is shown in Figure 7.7.

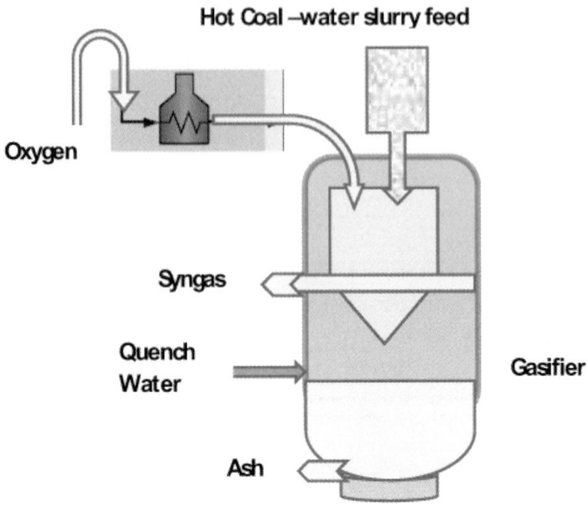

Fig. 7.7. Simplified diagram of Texaco Gasification process

In this process, hot coal and water slurry is fed to a pre-heater where water turns into steam before being fed to the gasifier under pressure from the top. From another inlet nozzle, pre-heated oxygen is injected such that it comes in contact with the coal at the top of the gasifier. Coal is gasified at the top at a temperature between 1000 °C -1300 °C and pressure of 2 to 8 MPa. The bottom of the gasifier is quenched with water. Syngas is taken out of the gasifier just above the quench section. Ash as molten slag flows down and is taken out of the gasifier section periodically. In a typical commercial gasification plant, 1,000 tons of coal produces about 1.5 million standard meters cubed (MMsm^3) of syngas[3].

The major contaminant in syngas is tar particles. It needs to be first cleaned to avoid environmental emission and is removed using cyclones. Then it is cleaned further of sulfur compounds, as shown above in the reaction mechanisms as they are highly toxic materials. Acid gas is removed by a scrubbing mechanism. If the syngas is to be used as petrochemical feedstock, then it needs rigorous purification.

7.7 Heavy oil economics

The ultimate objective of any refinery operation is the economics. It is true that one process or technology is not the solution. The economical solution significantly depends on the choice of technology, properties of the feedstock, and the quality of the final product. Heavier hydrocarbons are more difficult to upgrade and produce refractory materials especially from the delayed coking or FCC plants. Those materials require more severe conditions to process. Another critical issue involves environmental regulations. In order to meet the regulations, refiners need better catalysts, and higher pressure and higher temperature hydrogenation reactors, and at the same time need to justify this against overall refinery economics.

Lastly, the refinery economics is the most difficult thing to assess. It is highly site-specific and project-specific and depends on the cost of the feed, process conditions, final product purity, and its value. As shown in Figure 7.8, as the degree of upgrading and refining steps increase, the Capex and Opex both increase. However, based on the final product quality and demand, the profit margin increases as the product price increases. Nevertheless, the final profit margin depends on upgrading, refining, and transportation that are required for the final product to reach the market. Rate of return depends on the net revenue vs. Capex.

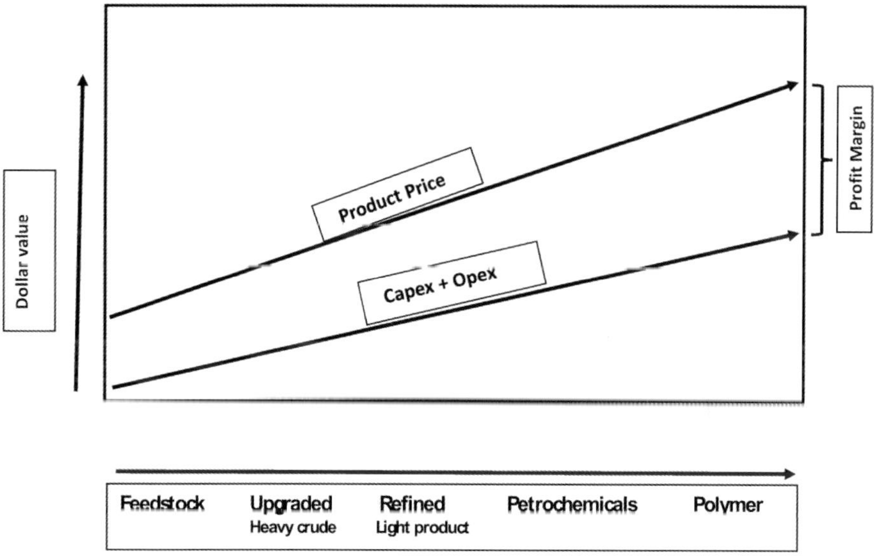

Fig. 7.8. Projected economics of various refining steps

References for Chapter 7

1. M. Volk, US patent #8,419,931.
2. J. G. Speight, "The Chemistry and Technology of Petroleum" Marcel Decker Publisher, New York (1980).
3. N. Berkowitz , Coal Technology, Academic Press, CA (1994).
4. R.R. Vezirov, Visbreaking Technology, *Chem & Tech of Fuel Oil,* Vol 46, No 6, (2011).
5. D.K.Banerjee, M. Volk, K. Wisecarver, G.V.Nava, US Patent application publication # US2017/0321127
6. Dwijen Banerjee, Amaka Waturuocha, Glixon Mavarez, Michael Volk; Internal communication prepared for publication, University of Tulsa (2018)

8

The future of hydrocarbon fuel

Recently, several underdeveloped countries in the world have progressed towards "developing" status with an increasing economic boom. China and India are the best examples; their economies have grown rapidly, and both now hold the status of being among the top ten economies of the world. With the global economy increasing, so does the demand for transportation fuels. It seems hydrocarbons will remain the primary source for transportation by road, by air, or by sea for a long time to come. Transportation fuel is increasingly becoming cleaner and almost the same policy is now adopted worldwide. Recently, the International Maritime Organization has introduced a new mandatory rule IMO-2020, to be obeyed by all shipping companies all over the world: a sulfur limitation of 0.5 weight percentage in bunker fuel starting year 2020.

International efforts to limit green house gases (GHG) contributing to global warming mean energy industries are looking for alternative paths. "Bio-fuels" have been in the market for some time now and will continue to be part of the transportation fuel market. In the US, 10% ethanol blend hydrocarbon fuel is available almost all over the country. Some countries such as Brazil have introduced E-85, meaning an 85% ethanol blend, and in some places, 100% ethanol fuel is available. Recent news from China indicates that they have introduced methanol-based fuel in heavy vehicles that are running with M85, i.e., 85% methanol fuel.

In the US, flexible-fuel vehicles are available that can run with pure hydrocarbons or alcohol. The automakers are introducing another alternative fuel in the market—compressed natural gas—popularly known as CNG. In highly congested places like New Delhi, in India, almost all local public transportation vehicles, including taxis, run with CNG.

There is an increasing demand for renewable fuel and heavy investments going on all over the world in search of renewable energy or alternative energy. Electricity generated by solar and wind power is getting the maximum attention. The cost of generating electricity using solar panels has been decreasing exponentially over the last couple of years.

Electrification of transportation is now the name of the game. Electric vehicles (EVs) are becoming popular but are far away from replacing liquid hydrocarbons. Elon Musk of Tesla is shooting for the trophy "EV king of the road." According to a recent report by the EIA's Global EV Outlook (2018), the circulation of electric vehicles in the US has increased by three-fold in the last five years, whereas the number is almost five-fold for the same period globally. European conventional car manufacturers are replacing their internal combustion engine cars with EV car models, and it is expected that EV cars may triple in the EU in the next five years. They are penetrating the EU market with more affordable cars for the common people, as the present models are now limited to only the people who can afford them.

At the same time, this has a profound effect on transportation fuel demand, which is expected to decrease continuously beyond 2020. However, electricity generation and supply are the main issue. If coal-fueled power plants are used, the carbon footprint is not decreasing as it is perceived by the general public. Power generation with any carbon-rich source is an energy- and GHG-intensive process. GHG is not only CO_2, it also includes methane (CH_4), a part of natural gas production, and nitrous oxide (N_2O). It is actually a big challenge to accurately calculate total GHG emissions, as most of the reported data do not include lifecycle-based models with details from various sources of emissions in every step of the process.

The only solution for introducing in a shorter period a massive amount of electricity into the market with zero GHG is nuclear power plants. This was on the radar screen for some time before the Fukushima nuclear disaster happened in Japan. This was not the fault of the nuclear reactor design, as an unexpected and unheard of natural disaster happened. But some countries started rolling back their proposals, and countries like Germany and Switzerland imposed a total ban on future nuclear plants. According to the recent discussions going on in the European Union, they are planning to go "Carbon-Neutral" by 2050.

But there is a transition period, as always in every sector of any industry. It seems that natural gas is the transitional fuel for the future to act as a low-carbon source, until there is any hope of replacing hydrocarbon with alternate fuel. The fact that natural gas emits the lowest GHG as compared to any other liquid or solid hydrocarbons suggests that gradual replacement is needed, instead of, as some argue, replacing hydrocarbons with alternate fuel immediately. Most importantly, if the power is not generated from a clean source, it does not matter how the power is used—it is not clean.

Although renewable energy is getting considerable attention from all over the world and is growing rapidly, it is nowhere near meeting the target of global energy demand. According to a recent report (2018) from the International Energy Agency (IEA), renewable energy may account for just below 15% of the total global energy demand by 2025 and may be responsible for one third of the global electricity produced. According to the same report, solar energy is expected to grow in commercial and industrial areas and, in the same period, wind energy to follow suit. The biggest growth will be in China and Brazil.

For the future, the global trend is increasing towards renewables, solar panels, wind power, fuel cells, and most importantly, EVs. It is true that the emergence of these technologies will steadily reduce the demand for hydrocarbon transportation fuels. During this period, the prospect of strong hydrocarbon growth is expected to shift towards petro-chemicals from the transportation area. Especially in North America, shale resources will force the refiners to switch from the production of gasoline and diesel towards petrochemicals in order to increase their profit margins to survive. This can only be achieved through higher integration of refining and petrochemical capacities, as discussed in earlier chapters, 5 and 6.

Nevertheless, it is also true that all those technologies combined will not be able to replace all hydrocarbons in the near future. However, we must keep our eyes open on the electrification of the transportation sector over the next couple of decades; the aggressive nature of new automobile industries, like Tesla, may make it possible.

According to the Bloomberg report shown in Figure 8.1, the capital cost of building a low-emission power plant with clean coal technology (90% carbon capture) costs almost five times that of a natural gas-based power plant and will cost more than double that of a solar panel power plant. Utility companies have been closing down old coal-based plants (not clean-coal) and replacing them with cheaper, lower-emitting natural gas. A recent 2018 EIA report suggests that in the US, on average, 50% of the emission reduction took place due to coal-to-gas switching, based on per unit electricity generated.

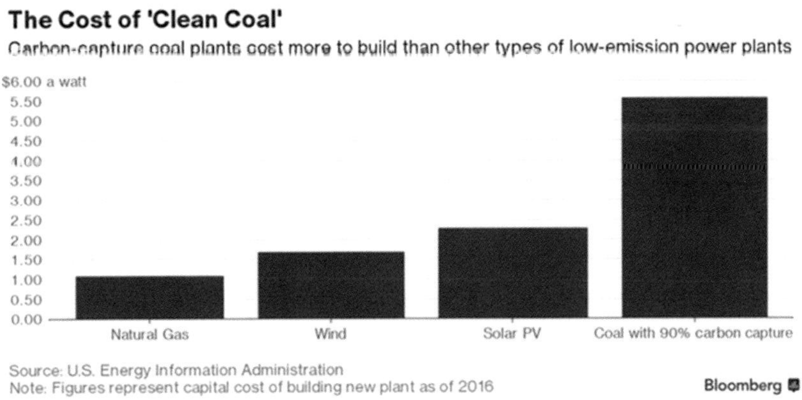

Fig. 8.1. The cost of a clean coal plant as compared to other renewable energy plants (source: Bloomberg and EIA)

Figure 8.2 shows that oil, gas, and coal still dominate and are expected to dominate as the main energy sources for years to come. The present prevailing assumption shared by most forecasters is that green energy sources—mostly wind and solar—will take a growing share of the world power supply. However, in order to increase that share, first they have to improve the technology further and the cost of production has to go down enough to be competitive with the hydrocarbon-based technologies. The only short-term way out of these dilemmas is via vast new nuclear power stations, certainly generating low-carbon electricity, but at an astronomic cost and producing many headaches regarding waste handling and disposal.

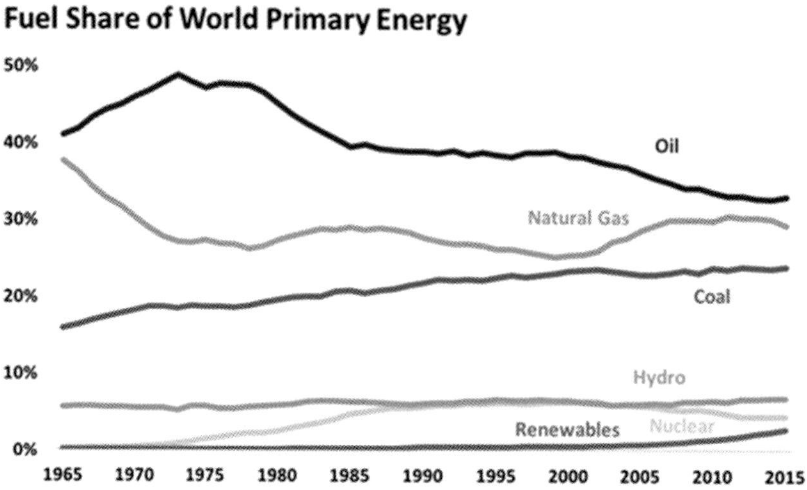

Fig. 8.2. Comparison data of the percentage share of various energy sources in the world (source: BP statistical review of world energy 2016)

Today, the world is in a constant state of change; it seems to be moving at the speed of light, whether it is about going to outer space and reaching places not thinkable before, or developing wireless technologies, or developing social media that brings people together. But nowhere is this more evident than in the energy sector, where the story of innovation is everywhere. We do not know what the transportation sector will look like, say, in the year 2040 or 2050.

The researchers have done an amazing job, and productive people have implemented the science properly and effectively. More work from researchers must be done by collaborating and sharing resources. I know from my own long research experience that we can benefit tremendously when we work as a team and share our wisdom from experts in their respective areas. Then comes leadership; the research team needs a good manager to make sure everyone has the resources they need and to make the atmosphere a productive place to work. As I come to the close, I would like to thank again the research communities all over the world whose dedication makes the production and operation a success. Everything the world is fighting for daily, to survive directly or indirectly, depends on energy. The quality of life is dependent on the availability of affordable energy, whether it is a rich country or a poor country.

Very recently, the US government has been showing less responsibility for investing in research, while other countries are investing heavily. It is true that the private companies are working hard, but the risk of research must be shared by all. We will continue to pursue new innovations, as people want changes for the better. I am from the baby-boomer older generation; millennials are now sitting in corner offices, with access to all available information at their fingertips at the speed of light, literally (of course with the help of a handheld mobile device), so it is not surprising that they will do a better job than that of earlier generations.

I would like to end this book with a quotation from the person (from the living generation) whom I admire the most and whom I follow: Bill Gates. He himself is a great admirer of Leonardo da Vinci and he wrote the following in Time magazine (February 2019 issue):

"What does a Renaissance artist have to do with optimism? For me, the connection is innovation. I feel optimistic about the future because I know that advances in human knowledge have improved life for billions of people, and I am confident they will keep doing so. And although I am no art expert, everything I have learned about Leonardo leads me to believe he was the most innovative thinker ever."

List of patents

The author as inventor / co-inventor

US Patents

1 #6,755,962
2 #7,922,895
3 #7,975,763
4 #8,365,823
5 #8,431,015
6 #8,462,789
7 #8,555,970
8 #8,689,865
9 #8,720,547
10 #8,720,548
11 #8,720,549
12 #8,720,550
13 #8,807,220
14 #8,905,127
15 #8,960,286

Canadian Patents

1- #2807850
2- #2777956
3- #2777942
4- #2777792
5- #2777947
6- #2777862
7- #2777859
8- #2704575
9- #2704689
10- #2678841
11- # 2026840

Index

A

B

C

D

E

F

G

H

J

K

L

M

N

O

P

R

S

T

U

V

W

X

Z